ELECTRICITY 4

ELECTRICITY

4

AC/DC MOTORS, CONTROLS AND MAINTENANCE

SIXTH EDITION

WALTER N. ALERICH
JEFF KELJIK

Delmar Publishers

I(T)P An International Thomson Publishing Company

Albany • Bonn • Boston • Cincinnati • Detroit • London • Madrid • Melbourne
Mexico City • New York • Pacific Grove • Paris • San Francisco • Singapore • Tokyo
Toronto • Washington

NOTICE TO THE READER

Delmar Staff:
Publisher: Susan Simpfenderfer
Acquisitions Editor: Paul Shepardson
Project Development Editor: Michelle Ruelos Cannistraci
Production Coordinator: Karen Smith
Marketing Manager: Lisa Reale

Copyright © 1996
By Delmar Publishers
a division of International Thomson Publishing Inc.
The ITP logo is a trademark under license

Printed in the United States of America

For more information, contact:
Delmar Publishers, Inc.
3 Columbia Circle, P.O. Box 15015
Albany, New York 12212-5015

International Thomson Publishing Europe
Berkshire House 168-173
High Holborn
London WC1V7AA
England

Thomas Nelson Australia
102 Dodds Street
South Melbourne, 3205
Victoria, Australia

Nelson Canada
120 Birchmount Road
Scarborough, Ontario
Canada MlK 5G4

Delmar Publishers' Online Services
To access Delmar on the World Wide Web, point your browser to:
http://www.delmar.com/delmar.html
To access through Gopher:
gopher://gopher.delmar.com (Delmar Online is part of the "thomson.com", an Internet site with information on more than 30 publishers of the International Thomson Publishing organization.)
For information on our products and services:
email:info@delmar.com
or call 800-347-7707

International Thomson Editores
Campos Eliseos 385, Piso 7
Cot Polanco
11560 Mexico D F Mexico

International Thomson Publishing GmbH
Königswinterer Strasse 418
53227 Bonn
Germany

International Thomson Publishing Asia
221 Henderson Road
#05-10 Henderson Building
Singapore 0315

International Thomson Publishing-Japan
Hirakawacho Kyowa Building, 3F
2-2-1 Hirakawacho
Chiyoda-ku, Tokyo 102
Japan

4 5 6 7 8 9 10 xxx 01 00 99

Library of Congress Cataloging-in-Publication Data
Alerich, Walter N.
    Electricity 4 : AC/DC motors, controls, and maintenance/Walter N. Alerich, Jeff Keljik. -- 6th ed.
      p.   cm.
      Includes index.
      ISBN 0-8273-6593-4 (pbk.)
      1. Electric motors, Alternating current. 2. Electric generators--Alternating current. 3. Electric controllers. I. Keljik, Jeff. II. Title.
TK2712.A55 1996
621.31'042--dc20
                              95-42327
                              CIP

CONTENTS

PREFACE

The sixth edition of Electricity four has been reorganized to provide more continuity of topics and better flow of concepts. It has been updated with new material and new artwork to better reflect the current work place. At the same time, the text has retained the features and style of previous editions that has made it so popular.

The text introduces the concepts of AC and DC motors, the associated controls and the maintenance of this equipment. The material is broken down into short segments which concentrate on specific concepts or application of particular types of equipment. The detailed explanations are written in easy to understand language which concisely presents the needed knowledge. There are many illustrations and photographs which help to provide technical understanding and provide real world reference. This type of explanation and application better prepares the student to perform effectively on the job in installation, troubleshooting, repair, and service of electrical motors and controls.

The knowledge obtained in this book permits the student to progress further in the study of electrical systems. It should be understood that the study of electricity and the application of electrical products are continually changing.

The electrical industry constantly introduces new and improved devices and material which in turn lead to changes in installation and operation of equipment. Electrical codes also change to reflect the industry needs. It is essential that the student continue to learn and update their knowledge of the current procedures and practices.

The text is easy to read and the units have been grouped by general subject. There are summaries of each unit which provide an opportunity to restate the most important topics of the unit. There are summaries of the units provided to re-emphasize topic groups.

Each unit begins with the learning objectives. An Achievement Review at the end of each unit provides an opportunity for the reader to check their understanding of the material in small increments before proceeding. The problems in the text require the use of simple algebra and the student should be familiar with the math before trying to solve the equations. It is also essential that the reader have a basic understanding of the fundamentals of electrical circuits and electrical concepts.

It is recommended that the most recent edition of the National Electrical Code (published by the National Fire Protection Association) be available for reference and use, as the learner applies this text. Applicable state and local codes and regulations should also be consulted when making the actual installations.

Features of the sixth edition include:

- Reorganization of topics into more related topics and associated concepts
- Updated photos and artwork to reflect current equipment and practices
- Content updated to the most recent electrical code
- More extensive information on solid state controls
- More extensive information on motor installations
- Summaries and achievement reviews at the end of each unit

A combined instructor's guide for ELECTRICITY 1 through ELECTRICITY 4 is available. The guide includes the answers to the achievement review and the summary review for each text. Additional test questions covering the content of the four texts are also included. Instructors may use these questions to devise additional tests to evaluate student learning.

ABOUT THE AUTHORS

Walter N. Alerich, BVE, MA, has an extensive background in electrical installation and education. As a journeyman wireman, he has many years of experience in the practical applications of electrical work. Mr. Alerich has also served as an instructor, supervisor and administrator of training programs, and is well-aware of the need for effective instruction in this field. A former department head of the Electrical-Mechanical Department of Los Angeles Trade-Technical College, Mr. Alerich has written extensively on the subject of electricity and motor controls. He presently serves as an international specialist/consultant in the field of electrical trades, developing curricula and designing training facilities. Mr. Alerich is also the author of ELECTRICITY 3 and ELECTRIC MOTOR CONTROL LABORATORY MANUAL, and the coauthor of INDUSTRIAL MOTOR CONTROL.

Jeff Keljik has been teaching at Dunwoody Institute for over 17 years and is the Department Head of Electrical, Electronics, and Computer Technology. As a licensed master electrician he is also in charge of electrical construction and maintenance at Dunwoody. Before beginning his career in education, he worked for five years as a maintenance electrician.

In addition to his teaching and administrative duties, Mr Keljik is former chairman of the Minnesota Technical, Trade and Industrial Association (Electrical Section). He is also a consultant with industry and serves as an electrical coordinator on several international projects.

ACKNOWLEDGMENTS

Grateful acknowledgement is given the following individuals for their contributions to this edition of ELECTRICITY 4:

Houston Baker
Auburn Electrical Construction Co., Inc.
Auburn, AL 36830

Allen Beiling
Assabet Valley Regional Vocational High School
Marlborough, MA 01752

Robert W. Blakely
Mississippi Gulf Coast Community College
Gulfport, MS 39507

Gerald Le Cardi
R. McKee Vocational and Technical High School

Larry A. Catron
Scott County Vocational School
Gate City, VA 24251

Keith DeMell

Glenn Graham
Matbaum AVTS
Philadelphia, PA 19134

John Moyer
Hodgson Vocational Technical High School
Newark, DE 19702

Thomas J. Ritchie, Jr.
Medford Vocational-Technical High School
Medford, MA 02155

Dave Welsch
Welsh & Sons Electric, Inc.
Niles, MI 49120

Don West
Boonslick Area Vo-Tech
Boonville, MO 65233

DEDICATION:

I would like to dedicate this sixth edition to my son, Jonathan. He has always understood when I have not been able to spend time with him because of my time dedicated to writing.

ELECTRICAL TRADES

The Delmar series of instructional material for electrical trades includes the texts, text-workbooks, laboratory manuals, and related information workbooks listed below. Each text features basic theory with practical applications and student involvement in hands-on activities.

ELECTRICITY 1   ELECTRICITY 3
ELECTRICITY 2   ELECTRICITY 4
ELECTRIC MOTOR CONTROL
ELECTRIC MOTOR CONTROL LABORATORY MANUAL
INDUSTRIAL MOTOR CONTROL
ALTERNATING CURRENT FUNDAMENTALS
DIRECT CURRENT FUNDAMENTALS
ELECTRICAL WIRING – RESIDENTIAL
ELECTRICAL WIRING – COMMERCIAL
ELECTRICAL WIRING – INDUSTRIAL
PRACTICAL PROBLEMS IN MATHEMATICS FOR ELECTRICIANS

EQUATIONS BASED ON OHM'S LAW

P = Power in watts
I = Intensity of Current in Amperes
R = Resistance in Ohms
E = Electromotive Force in Volts

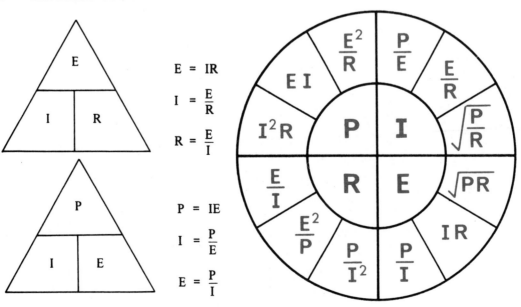

$$E = IR$$

$$I = \frac{E}{R}$$

$$R = \frac{E}{I}$$

$$P = IE$$

$$I = \frac{P}{E}$$

$$E = \frac{P}{I}$$

THE DC SHUNT MOTOR

OBJECTIVES

After studying this unit, the student will be able to

- list the parts of a dc shunt motor.

- draw the connection diagrams for series shunt and compound motors.

- define torque and tell what factors affect the torque of a dc shunt motor.

- describe counter emf and its effects on current input.

- describe the effects of an increased load on armature current, torque, and speed of a dc shunt motor.

- list the speed control, torque, and speed regulation characteristics of a dc shunt motor.

- make dc motor connections.

The production of electrical energy, and its conversion to mechanical energy in electric motors of all types, is the basis of our industrial structure. Dc motor principles are given in ELECTRICITY 1.

CONSTRUCTION FEATURES

Dc motors closely resemble dc generators in construction features. In fact, it is difficult to identify them by appearance only. A motor has the same two main parts as a generator – the field structure and the armature assembly consisting of the armature core, armature winding, commutator, and brushes. Some general features of a dc motor are shown in figure 1–1A and B.

The Field Structure

The field structure of a motor has at least two pairs of field poles, although motors with four pairs of field poles are also used (figure 1–2A). A strong magnetic field is provided by the field windings of the individual field poles. The magnetic polarity of the field system is arranged so that the polarity of any particular field pole is opposite to that of the poles adjacent to it.

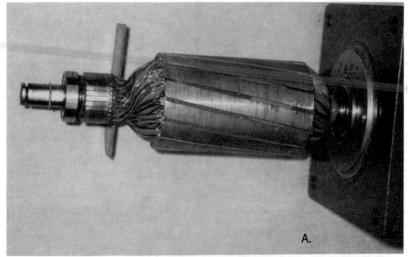

A.

CARBON BRUSH CONNECTIONS
TO COMMUTATOR BARS

B.

Fig. 1–1   A) DC motor armature with commutator bars   B) Permanent
magnet DC motor with rotor and carbon brush connections

The Armature

The armature of a motor is a cylindrical iron structure mounted directly on the motor shaft (figure 1–2B). In DC motors, the armature is the rotating component of the motor. Armature windings are embedded in slots in the surface of the armature and terminate in segments of the commutator. Current is fed to these windings on the rotating armature by carbon brushes which press against the commutator segments. This current in the armature sets up a magnetic field in the armature which acts with the magnetic field of the field

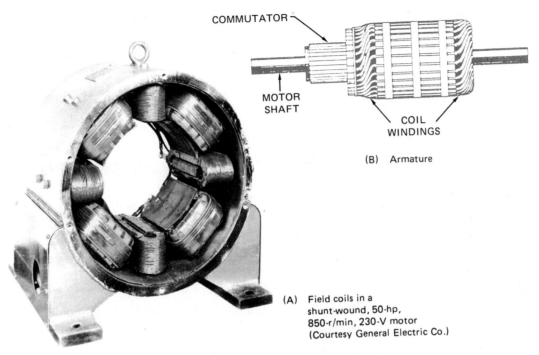

(A) Field coils in a
shunt-wound, 50-hp,
850-r/min, 230-V motor
(Courtesy General Electric Co.)

(B) Armature

Fig. 1–2 Field structure and armature assembly of a motor

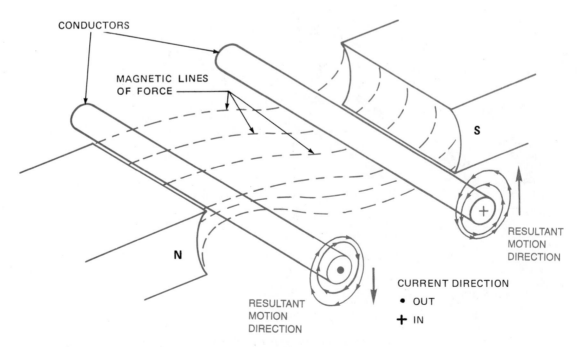

Fig. 1–3 Torque, or force direction on a current-carrying conductor in a magnetic field

poles. These magnetic effects are used to develop torque which causes the armature to turn (figure 1–3). The commutator changes the direction of the current in the armature conductors as they pass across poles of opposite magnetic polarity. Continuous rotation in one direction results from these reversals in the armature current.

Figure 1–4 is a cutaway view of a dc motor available with horsepower ratings ranging from 25.0 hp to 1,000 hp.

TYPES OF DC MOTORS

Shunt, series, compound and permanent magnet motors are all widely used. The schematic diagram for each type of motor is shown in figure 1–5. The selection of the type of motor to use is based on the mechanical requirements of the applied load. A shunt motor has the field circuit connected in shunt (parallel) with the armature, while a series motor has the armature and field circuits in series. A compound motor has both a shunt and a series field winding. A permanent magnet motor only has armature connections.

MOTOR RATINGS

DC motors are rated by their voltage, current, speed, and horsepower output.

TORQUE

The rotating force at the motor shaft produced by the interaction of the magnetic fields of the armature and the field poles is called *torque*. The magnitude of the torque

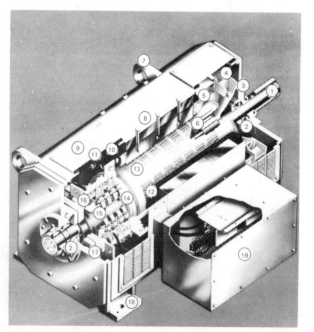

1. Main shaft
2. Bearings
3. Grease "meter"
4. Ventilating fan
5. Armature banding
6. Armature equalizer coil assembly
7. Lifting lugs
8. Frame
9. Inspection plate
10. Main field coil
11. Commutating coils
12. Main field coil
13. Armature
14. Commutator connections to armature turns
15. Commutator
16. Brushholder
17. Brushholder yoke
18. Mounting feet
19. Terminal conduit box

Fig. 1–4   Assembled 25-hp dc motor (*Courtesy of General Electric, DC Motor and Generator Department*)

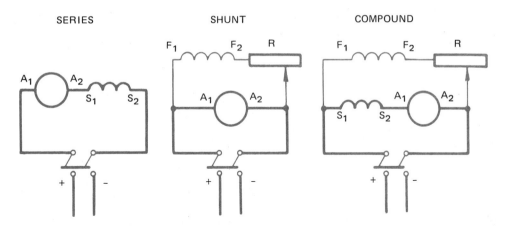

Fig. 1–5   Motor field connections

increases as the twisting force of the shaft increases. Torque is defined as the product of the force in pounds and the radius of the shaft or pulley in feet.

For example, a motor which produces a tangential force of 120 pounds at the surface of the shaft 2 inches in diameter or 1 inch radius, has a torque of 10 foot-pounds (ft-lb).

$$\text{Torque} = \text{Force} \times \text{Radius}$$

$$= 120 \times 1/12 = 10 \text{ ft-lb}$$

Torque in a motor depends on the magnetic strengths of the field and the armature. Since the armature field depends on armature current, the torque increases as the armature current, and consequently the strength of the armature magnetic field, increase.

It is necessary to distinguish between the torque developed by a motor when operating at its rated speed and the torque developed at the instant the motor starts. Certain types of motors have high torque at rated speed but poor starting torque. The many types of loads which can be applied to motors mean that the torque characteristic must be considered when selecting a motor for a particular installation.

STARTING CURRENT AND COUNTER ELECTROMOTIVE FORCE

The starting current of a dc motor is much higher than the running current while the motor is operating at its rated speed. At the instant power is applied, the armature is motionless and the armature current is limited only by the very low armature circuit resistance. As the motor builds up to its rated speed, the current input decreases until the motor reaches its rated speed. At this point, the armature current stops decreasing and remains constant.

Factors other than armature resistance also limit the current. Figure 1–6 illustrates a demonstration which shows the "generator" action within a motor that accounts for the decrease in current with a speed increase.

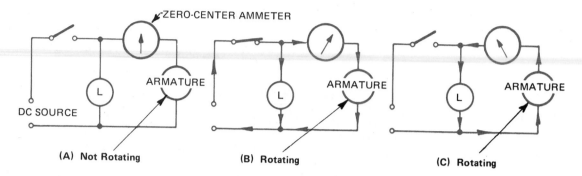

Fig. 1–6 Demonstration of counter emf

In figure 1–6, a dc motor and a lamp (each with the same voltage rating) are connected in parallel to the dc source. A zero-center ammeter connected in the circuit indicates the amount and direction of the current to the motor. When the line switch is open (A), there is no current in any part of the circuit. When the switch is closed (B), the lamp lights instantly and the ammeter registers high current to the motor. The motor current decreases as the motor speed increases and remains constant when the motor reaches its rated speed. The instant the switch is opened, the ammeter deflection reverses. The lamp continues to light but grows dimmer as the motor speed falls.

Two conclusions can be made from this demonstration:

1. A dc motor develops an induced voltage while rotating.

2. The direction of the induced voltage is opposite to that of the applied voltage and for this reason is called *counter emf.*

As the torque, or twisting effort, rotates the armature, the conductor coils of the armature cut the main field magnetic flux, as in a generator. This action induces a voltage into the armature windings which opposes line voltage.

The production of counter emf in a dc motor accounts for the changes in current to a motor armature at different speeds. When there is no current in the circuit, the motor armature is motionless and the counter emf is zero. The starting current is very high because only the ohmic resistance of the armature limits the current. As the armature starts to rotate, the counter emf increases and the line current decreases. When the speed stops increasing, the, value of the counter emf approaches the value of the applied voltage, but is never equal to it. The value of the voltage which actually forces current through the motor is equal to the difference between the applied voltage and the counter emf. At rated speed, this voltage differential will just maintain the motor at constant speed (figure 1–7A).

When a mechanical load is then applied to the motor shaft, both the speed and counter emf decrease. However, the voltage *differential increases* and causes an increase

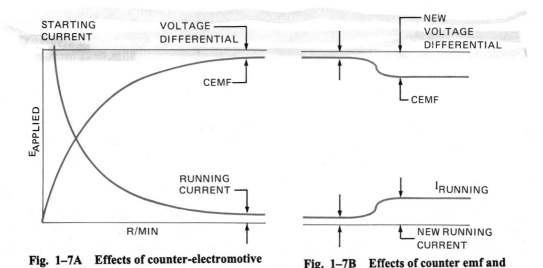

Fig. 1–7A **Effects of counter-electromotive force on the armature current**

Fig. 1–7B **Effects of counter emf and $I_{armature}$ when the load is increased**

of input current to the motor. Any further increase in mechanical load produces a proportional increase in input current (figure 1–7B).

The increase in motor current due to an increase in mechanical load also can be explained in terms of the torque. Since torque depends upon the strength of the magnetic field of the armature which, in turn, depends upon the armature current, any increase in mechanical load would require an increase in the armature current.

Since the starting current may be many times greater than the rated current under full load, large dc motors must not be connected directly to the power line for startup. The heavy current surges produce excessive line voltage drops which may damage the motor. The maximum branch-circuit fuse size for any dc motor is based on the fullload running current of the motor. Therefore, starters for dc motors generally limit the starting current to 150% of the full-load running current.

ARMATURE REACTION

Armature reaction occurs in dc motors and is caused by the stator magnetic field being distorted, or altered, in reaction to the armature magnetic field. The armature reaction is actually a bending of the motor magnetic field so that the brushes are no longer aligned with the neutral magnetic plane of the motor. If the brushes are not in alignment with this magnetic plane, the current conducted to the armature does not split equally in the armature conductors and therefore causes a voltage difference at the brushes. This causes sparking where the brush meets the commutator. In a motor with a constant load, the brushes can be shifted back into the neutral plane to reduce sparking. The brushes are shifted in the direction opposite to rotation. If the motor has a varying load, the neutral plane will constantly be shifting. To counteract the effects of the field distortion, some motors are designed with *interpoles* or *commutating poles*. These poles are connected in series with the armature circuit. Every change in armature current that would tend to distort the magnetic field is counteracted by the interpole magnetic field. See figure 1–3.

ROTATION

The direction of armature rotation of a dc motor depends on the direction of the current in the field circuit and the armature circuit (figure 1–8A). To reverse the direction of rotation, the current direction in *either* the field or the armature must be reversed. Reversing the power leads does not reverse the direction of armature rotation because this situation causes *both* the field and armature currents to become reversed as shown in figure 1–8B. To determine the direction of conductor movement, use the right-hand rule for motors. Use the right-hand as shown in figure 1–8C. The first finger indicates the direction of the flux (north to south), the center finger indicates the direction of the current flow (negative to positive) and the thumb will indicate the direction of the resultant thrust.

SPEED CONTROL AND SPEED REGULATION

The terms speed control and speed regulation should not be used interchangeably. The meaning of each is entirely different. *Speed regulation* refers to a motor's ability to maintain a certain speed under varying mechanical loads from no load to full load. It is expressed as a percent. The formula used is:

$$\% \text{ speed regulation} = \frac{\text{No load speed} - \text{Full load speed}}{\text{Full Load speed}} \times 100$$

Using this formula, we can determine that a motor that holds a constant speed between no load and full load has a 0% speed regulation.

Speed control refers to changing the motor speed intentionally by means of external control devices. This is done in a variety of ways and is not a result of the design of the motor.

Speed Control

DC motors are operated below normal speed by reducing the voltage applied to the armature circuit. Resistors connected in series with the armature may be used for voltage reduction. When the armature voltage is reduced while keeping the field current constant, the counter emf is too high. Therefore, the motor slows down to reduce the counter emf (figure 1–9A). The speed of a dc motor can also be brought below its rated speed by varying the voltage applied to the whole motor. However, this method is not used because there is a loss of torque along with the reduction in speed.

A dc motor may be operated above its rated speed by reducing the strength of the field flux. A rheostat placed in the field circuit varies the field circuit resistance, the field current and, in turn, the field flux.

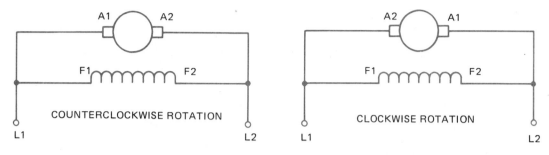

Fig. 1–8A Standard connections for shunt motors (From Herman/Alerich, *Industrial Motor Control,* **2nd Ed, copyright 1990 by Delmar Publishers Inc.)**

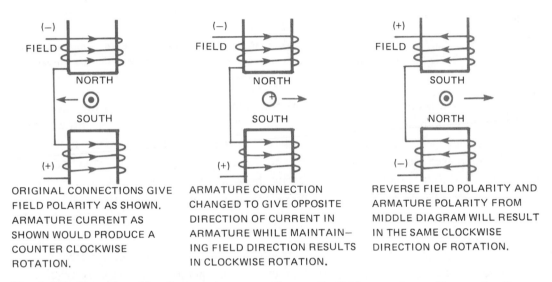

ORIGINAL CONNECTIONS GIVE FIELD POLARITY AS SHOWN. ARMATURE CURRENT AS SHOWN WOULD PRODUCE A COUNTER CLOCKWISE ROTATION.

ARMATURE CONNECTION CHANGED TO GIVE OPPOSITE DIRECTION OF CURRENT IN ARMATURE WHILE MAINTAIN— ING FIELD DIRECTION RESULTS IN CLOCKWISE ROTATION.

REVERSE FIELD POLARITY AND ARMATURE POLARITY FROM MIDDLE DIAGRAM WILL RESULT IN THE SAME CLOCKWISE DIRECTION OF ROTATION.

Fig. 1–8B Reversing either the armature connections or the field connections will cause the direction of armature rotation to change; changing both connections will result in the same direction of rotation

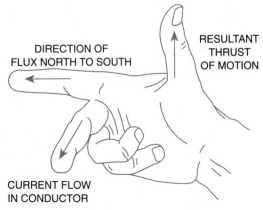

Fig. 1–8C Right-hand rule for motors using electron flow (From Keljik, *Electric Motors and Motor Controls,* **copyright 1995 by Delmar Publishers)**

Right-hand rule for motors using electron flow.

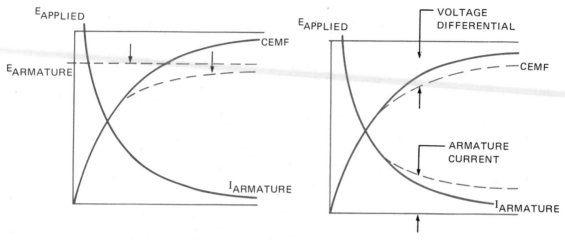

Fig. 1–9A   To reduce speed, reduce the armature voltage while keeping the field current constant

Fig. 1–9B   To increase speed, reduce the field current while keeping the armature voltage constant

Although it seems reasonable that a reduction in field flux reduces the speed, the speed actually increases because the reduction of flux reduces the counter emf and permits the applied voltage to increase the armature current. The speed continues to increase until the increased torque is balanced by the opposing torque of the mechanical load. When the field flux is reduced while keeping the armature voltage constant, the counter ernf in the armature drops. As a result, there is a larger voltage differential which causes an increase in armature current. This develops more torque to increase the speed of the motor (figure 1–9B).

Caution: Since motor speed increases with a decrease in field flux, the field circuit of a motor should never be opened when the motor is operating, particularly when it is running freely without a load. An open field may cause the motor to rotate at speeds that are dangerous to both the machine and to the personnel operating it. For this reason, some motors are protected against excessive speed by a field rheostat which has a nofield release feature. This device disconnects the motor from the power source if the field circuit opens.

THE SHUNT MOTOR

Two factors are important in the selection of a motor for a particular application: (1) the variation of the speed with a change in load, and (2) the variation of the torque with a change in load. A shunt motor is basically a constant speed device. If a load is applied, the motor *tends* to slow down. The slight loss in speed reduces the counter emf and results in an increase of the armature current. This action continues until the increased current produces enough torque to meet the demands of the increased load. As a result, the shunt motor is in a state of stable equilibrium because a change of load always produces a reaction that adapts the power input to the change in load.

The basic circuit for a shunt motor is shown in figure 1–10A. Note that only a shunt field winding is shown. Figure 1–10B shows the addition of a series winding to counteract the effects of armature reaction. From the standpoint of a schematic diagram, figure 1–10B represents a compound motor. However, this type of motor is not considered to be a compound motor because the commutating winding is not wound on the same pole as the field winding and the series field has only a few turns of wire in series with the armature circuit. As a result, the operating characteristics are those of a shunt motor. This is so noted on the nameplate of the motor by the terms compensated shunt motor or stabilized shunt motor.

Speed Control

A dc shunt motor has excellent speed control. To operate the motor above its rated speed, a field rheostat is used to reduce the field current and field flux. To operate below rated speed, reduce the voltage applied to the armature circuit.

A more modern method of speed control is the electronic speed control system. The principles of control are the same as the manual controls. Speeds above normal are achieved by reducing the field voltage electronically and speeds below normal reduce the voltage applied to the armature.

Rotation

The direction of armature rotation may be changed by reversing the direction of current in either the field circuit or the armature circuit. For a motor with a simple shunt field circuit, it may be easier to reverse the field circuit lead. If the motor has a series winding, or an interpole winding to counteract armature reaction, the same *relative* direction of current must be maintained in the shunt and series windings. For this reason, it is always easier to reverse the direction of the armature current.

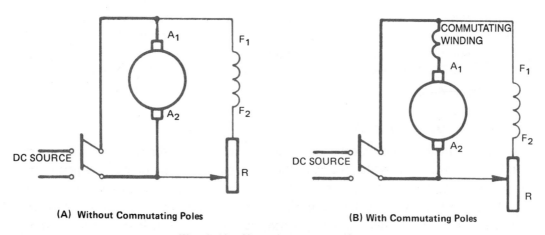

(A) Without Commutating Poles   (B) With Commutating Poles

Fig. 1–10   Shunt motor connections

Torque

A dc shunt motor has high torque at any speed. At startup, a dc shunt motor develops 150 percent of its rated torque if the resistors used in the starting mechanism are capable of withstanding the heating effects of the current. For very short periods of time, the motor can develop 350 percent of full load torque, if necessary.

Speed Regulation

The speed regulation of a shunt motor drops from 5 percent to 10 percent from the no-load state to full load. As a result, a shunt motor is superior to the series dc motor, but is inferior to a compound-wound dc motor. Figure 1–11A shows a dc motor with horsepower ratings ranging from 1 hp to 5 hp.

PERMANENT MAGNET MOTORS

A variation on the dc shunt motor principle is the PM (Permanent Magnet) motor. Two varieties are available. One style of PM motor uses a permanently magnetized material such as Alnico or ceramic magnets mounted in the stator to provide a constant magnetic field. The rotor is supplied with dc through a brush and commutator system. The result is similar to a dc shunt-type motor, but it has a very linear speed/torque curve.

Another type of PM motor uses the permanent magnets mounted in the rotor. Because dc is still supplied to the motor, commutation must be provided to properly magnetize the stator in relation to the rotor, to provide rotational torque. The commutator segments are actually connected to the stator windings and a set of sliding contacts on the rotor provides the proper electrical connection from the dc source to the proper commutator segments on the stator. This type of PM motor can be produced in larger-horsepower models than the PM stator types. PM motors are generally smaller than 5 hp. Figure 1–11B compares physical size of PM to shunt motor.

If the motors are not providing hp or torque, the problem could be that the magnets have lost some of the original magnetic strength. Another problem that can occur is the demagnetization of the permanent magnet material. This can happen when the motors are running in one direction and then quickly reversed under power. Some control circuits provide protection from quick reversals; others compensate for this problem by applying a small voltage during reversing.

BRUSHLESS DC MOTORS

Instead of using mechanical commutation to supply a field and power to the rotor, the use of electronics to switch the stator field can be used. The rotor uses a permanent magnet so that no direct power is supplied to the rotor. In order to switch the power supply to the field windings, sensing devices must be used to determine rotor movement. As the rotor speed increases or decreases, the sensor relays the information to the electronic

Fig. 1–11A   Direct current motor, 1hp to 5hp (*Courtesy of General Electric, DC Motor and Generator Department*)

Fig. 1–11B   Comparison of D.C. permanent magnet motor and wound field motor. (*Courtesy Bodine Electric Company***. From Keljik,** *Electric Motors and Motor Controls***, copyright 1995 by Delmar Publishers)**

switching supply. The electronic supply constantly adjusts to provide the proper level of voltage to the proper stator poles to maintain speed and direction. See figure 1–12.

STEPPING MOTORS

Another type of motor that can use a permanent magnet on the rotor is called *a stepping motor*. Instead of having a continuous supply of power and a continuous rotation, the

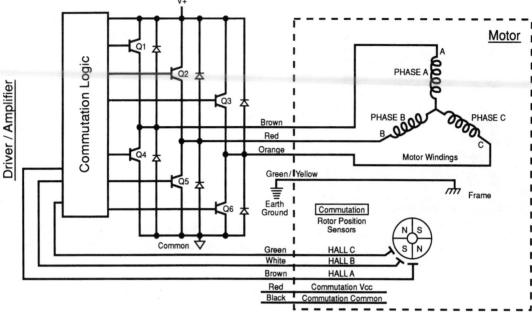

Fig. 1–12 Brushless DC motor control schematic (*Courtesy Bodine Electric Company*. From Keljik, *Electric Motors and Motor Controls*, copyright 1995 by Delmar Publishers)

rotor moves in steps as the stator is energized. The advantage of this type of motor is that motion can be monitored and exact degrees of rotation can be obtained from the input to the motor. These motors do not produce a great deal of torque so they are often used in small equipment needing incremental motion or "motion in steps."

The concept of the motor is to energize the stator field and allow the rotor poles to move into a desired position that provides magnetic alignment. Instead of providing a rotating magnetic field or a dc field with commutation, the fields are more stationary. As seen in figure 1–13, the stator can be energized by moving switches 1 and 2 to either position A or B. The first switch sequence shown in figure 1–13 will result in clockwise rotation; the second sequence produces counterclockwise rotation.

A simple stepper motor concept is explained using a permanent magnet on the rotor with just two sets of poles on the stator. Actually the rotor is made up of many magnetic poles aligned with "teeth" on the rotor. These teeth are spaced so that only one set of teeth are in perfect alignment with the stator poles at any one time. If we take the number of times that stator power must be applied to move one tooth through 360° of rotation, we can compute the step angle. For example, if the tooth moves 360° with 200 steps of power (application of stator power), the step angle is calculated by dividing the 360° by 200; this gives us 1.8° of motion per step. The step angle will determine how fine the steps of motion are for a given motor.

Other types of stepper motors use a high-permeability rotor instead of a permanent magnet rotor. The rotor magnetic fields will align themselves and retain the magnetism while in operation. These stepper motors are called *variable-reluctance* stepping motors.

Most stepping motors use instructions or commands that are produced by computer processors. The step commands are generated to produce a desired motion and fed to an electronic controller board, then power is applied to the motor leads. (Figure 1–14).

SUMMARY

The dc shunt motor uses the shunt field as the main magnetic field in the stator. The shunt field is made up of many turns of small wire and is connected or *shunted* across the armature. The shunt field may have a series-connected rheostat to control the amount of current to the field. The principle of the dc motor relies on the concept of commutation. This commutator and brush connection always keeps the direction of the current and the direction of the magnetic field consistent. The speed and the current to the rotor are inversely proportional. If the rotor is spinning faster, there is more counter emf (CEMF) produced and less voltage differential and therefore less current. DC motors are used in a

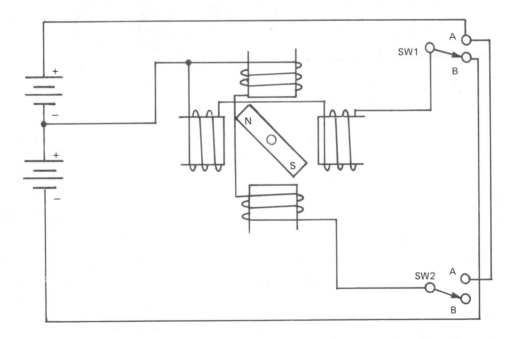

	CLOCKWISE ROTATION		COUNTERCLOCKWISE ROTATION	
	SWITCH 1	SWITCH 2	SWITCH 1	SWITCH 2
STEP 1	A	A	A	A
STEP 2	B	A	A	B
STEP 3	B	B	B	B
STEP 4	A	B	B	A
STEP 5	A	A	A	A

Fig. 1–13 Diagram illustrating how switching sequence produces steps of motion in a stepping motor

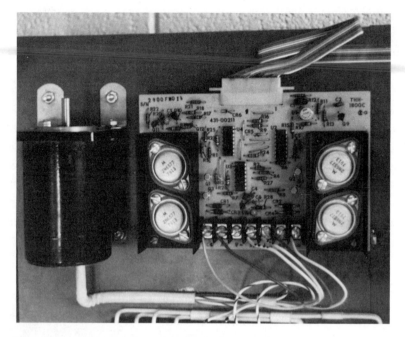

Fig. 1–14   Stepper motor and associated controller board. Note small
size of motor (only 65 oz. - in. of torque)

variety of styles for different purposes. There are many variations of the shunt motor used
in specialized purposes.

ACHIEVEMENT REVIEW

A.   Select the correct answer for each of the following statements and place the corresponding letter in the space provided.

1.   Dc motors are rated in   _____
   a. voltage, frequency, current, and speed.
   b. voltage, current, speed, and torque.
   c. voltage, current, and horsepower.
   d. voltage, current, speed, and horsepower.

2.   The generator effect in a motor produces a   _____
   a. high power factor.
   b. high resistance.
   c. counter electromotive force.
   d. reduced line voltage.

3.   A dc motor draws more current with a mechanical load
   applied to its shaft because the   _____

a. counter emf is reduced with the speed.
b. voltage differential decreases.
c. applied voltage decreases.
d. torque depends on the magnetic strength.

4. The direction of rotation of a compound interpole motor may
   be reversed by reversing the direction of current flow through the _____
   a. armature.
   b. armature or field circuit.
   c. armature, interpoles, and series field.
   d. shunt field.

5. The speed of a dc motor may be reduced below its rated speed
   without losing torque by reducing the voltage at the _____
   a. motor.
   b. series field.
   c. armature.
   d. armature and field.

6. Advantages of dc motors are _____
   a. simplicity in construction.
   b. speed control above and below base speed.
   c. excellent torque and speed control.
   d. horsepower for size.

B. Complete the following statements.

7. The twisting force exerted on the shaft of a motor is called _____and is due
   to the magnetic field interaction of the _____ and _____.

8. Field interpoles connected in series with the armature circuit of a motor help counteract the effects of_____ .

9. As a dc motor comes up to its rated speed, its armature current (decreases, remains
   the same, increases). (Underline the answer.)

10. The main factor controlling the armature current of a dc shunt motor operating at
    rated speed is the _____ .

U•N•I•T

2

THE DC SERIES MOTOR

OBJECTIVES

After studying this unit, the student will be able to

- draw the basic connection circuit of a series dc motor.
- describe the effects on the torque and speed of a change in current.
- describe the effects of a reduction of a load on the speed of a dc series motor,
- connect a dc series motor.

Despite the wide use of alternating current for power generation and transmission, the dc series motor is often used as a starter motor in automobiles and aircraft. This type of motor is also used as a traction motor because of its ability to produce a high torque with only a moderate increase in power at reduced speed.

The basic circuit for the series motor is shown in figure 2-1. The field circuit has comparatively few turns of wire of a size that will permit it to carry the full-load current of the motor.

TORQUE

A series motor develops up to 500 percent of its full load torque at starting. Therefore, this type of motor is used for railway installations, cranes, and other applications for which the starting load is heavy. The series motor is used in electric locomotives and are used for the drive wheels.

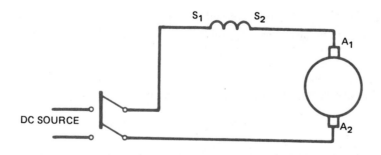

Fig. 2–1 Series motor connections

It should be remembered that the shunt motor operates at constant speed. In a shunt motor, any increase in torque requires a proportionate increase in armature current. In a series motor, the field is operated below saturation and any increase in load causes an increase of current in both the field and armature circuits. As a result, the armature flux and the field flux increase together. Since torque depends on the interaction of the armature and field fluxes, the torque increases as the square of the value of the current increases. Therefore, a series motor produces a greater torque than a shunt motor for the same increase in current. The series motor, however, shows a greater reduction in speed as mechanical load is added. A light load has little current draw and the armature and field current are reduced.

SPEED CONTROL AND SPEED REGULATION

The speed regulation of a series motor is inherently poorer than that of a shunt motor. If the mechanical load is reduced, a simultaneous reduction of current occurs in both the field and the armature. The reduction in the field current reduces the counter emf and the motor speeds up trying to rebuild the counter emf resulting from the reduced field flux. As a result, there is a greater increase in speed than would occur in a shunt motor for the same load change. If the mechanical load is removed entirely, the speed increases without limit and destruction of the armature through centrifugal force is certain to occur. For this reason, series motors are always permanently connected to their load.

If the maximum branch-circuit fuse size for any dc motor is limited to 150 percent of the full-load running current of the motor, the starters used with such motors must limit the starting current to 150 percent of the full-load current rating. Such starters must be equipped with an automatic, no-load release to prevent the armature from reaching dangerous speeds. The no-load release is set to open the circuit at the armature current corresponding to the maximum speed rating.

The speed of a series motor is controlled by varying the applied voltage. A series motor controller usually is designed to start, stop, reverse, and regulate the speed.

ROTATION

The direction of rotation may be reversed by changing the direction of the current either in the series field or the armature (figure 2-2).

MOTOR RATINGS

Series dc motors are rated for voltage, current, horse power, and maximum speed.

SUMMARY

The dc series motor has very high starting torque at very low speed. This characteristic makes it ideal for traction motors. These motors are used in fork lifts or diesel electric

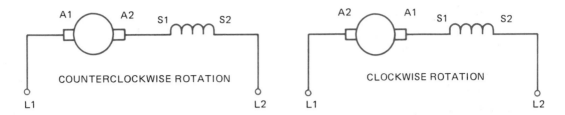

Fig. 2–2   Standard connections for series motors (From Herman/Alerich, *Industrial Motor Control*, 2nd Ed., copyright 1990 by Delmar Publishers Inc.)

locomotives. The relative speed of the motor is controlled by adjusting the applied voltage to the series field and the armature. The motor can be reversed by either changing the direction of current in the series field or the armature.

ACHIEVEMENT REVIEW

Select the correct answer for each of the following statements and place the corresponding letter in the space provided.

1. The torque of a series motor   _____
    a. is lower in its starting value than the starting torque for a
       shunt motor of the same horsepower rating.
    b. depends on the flux of the armature only.
    c. increases directly as the square of the current increase.
    d. increases with a load increase, but causes less of a reduc-
       tion in speed than a shunt motor for the same current increase.

2. For a series motor,   _____
    a. the field is operated below saturation.
    b. an increase in both the armature current and the field
       current occurs because of an increase in load.
    c. the reduction in speed due to an increase in load is
       greater than in the shunt motor.
    d. all of these.

3. Since a dc series motor has poor speed regulation,   _____
    a. a reduction in load causes an increase of current in both
       the field and armature.
    b. the removal of the mechanical load will cause the speed
       to increase without limit resulting in the destruction of
       the armature.
    c. it should not be connected permanently to its load.
    d. it does not require speed control.

4. The speed control for a dc series motor   _____
    a. is accomplished using a diverter rheostat across the series field.
    b. has an automatic no-field release feature included on all
       starters regardless of the limitations on the starting current.
    c. varies with the applied voltage.
    d. all of these.

5. A series motor controller usually is designed for   _____
    a. cranes.
    b. railway propulsion.
    c. starting loads when heavy.
    d. all of these.

6. Complete the electrical connections for the series motor.

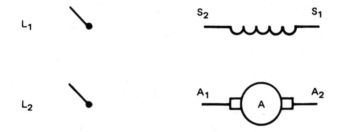

U • N • I • T

3

DC COMPOUND MOTORS

OBJECTIVES

After studying this unit, the student will be able to

- describe the torque, speed, rotation, and speed regulation and control characteristics of a cumulative compound-wound dc motor.
- perform the preliminary test for the proper installation of a cumulative compound motor.
- connect dc compound motors.
- describe the characteristics of a differential compound-wound dc motor.
- describe the characteristics of a cumulative compound-wound dc motor.

Compound-wound motors are used whenever it is necessary to obtain a speed regulation and torque characteristic not obtainable with either a shunt or a series motor. Since many drives need a fairly high starting torque and a constant speed under load, the compound wound motor is suitable for these applications. Some of the industrial applications include drives for passenger and freight elevators, stamping presses, rolling mills, and metal shears.

The compound motor has a normal shunt winding and a series winding on each field pole. As in the compound-wound dc generator, the series and shunt windings may be connected in long shunt (figure 3–1A), or short shunt (figure 3–1B).

When the series winding is connected to aid the shunt winding, the machine is a *cumulative compound motor*. When the series field opposes the shunt field, the machine is a *differential compound motor*. Using Fleming's right-hand rule for electromagnets, it can be seen that the two windings will either reinforce each other or try to cancel each other (figure 3–2).

TORQUE

The operating characteristics of a cumulative compound-wound motor are a combination of those of the series motor and the shunt motor. When a load is applied, the increasing current through the series winding increases the field flux. As a result, the torque for a given current is greater than it would be for a shunt motor. However, this flux increase causes the speed to decrease to a lower value than in a shunt motor. A cumulative

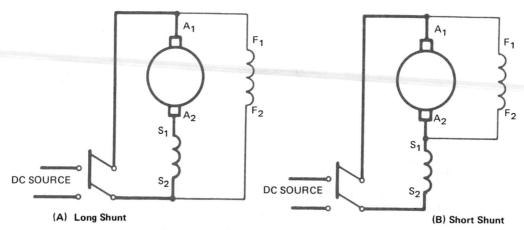

(A) Long Shunt   (B) Short Shunt

Fig. 3–1   Motor field connections

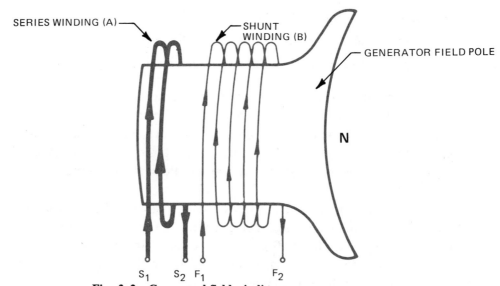

Fig. 3–2   Compound field windings

compound-wound motor develops a high torque with any sudden increase of load. It is best suited for operating varying load machines such as punch presses.

SPEED

Unlike a series motor, the cumulative compound motor has a definite no-load speed and will not build up to destructive speeds if the load is removed.

Speed Control

The speed of a cumulative compound motor can be controlled by the use of resistors in the armature circuit to reduce the applied voltage. When the motor is to be used for installations where the rotation must be reversed frequently, such as in elevators, hoists, and railways, the controller used should have voltage dropping resistors and switching arrangements to accomplish reversal.

Electronic Speed Control

A block diagram approach to the electronic speed control of a dc motor is presented in figure 3–3. The ac line is rectified by a full wave bridge to supply pulsating dc to the shunt field at a steady value. The dc supplied to the armature and the series field is controlled by an SCR (silicon controlled rectifier). By adjusting the firing time of the SCR, either more or less of the DC voltage available can be applied to the armature. If a low amount of dc voltage is applied, the torque is low and the resultant speed is low. If the SCR is fired early in the waveform and allowed to conduct for most of the cycle, then a larger amount of voltage is applied to the armature, more current results, and torque increases to spin the armature at a higher speed. Speeds below rated motor speed are produced by lowering the amount of voltage applied to the armature while keeping the shunt field steady; if higher-than-normal speed is needed, the shunt field can be weakened. Most speed controls have some type of feedback to sense the armature current, compare it to the set speed, and adjust the SCR firing angle to compensate for varying mechanical loads to keep the motor speed regulated.

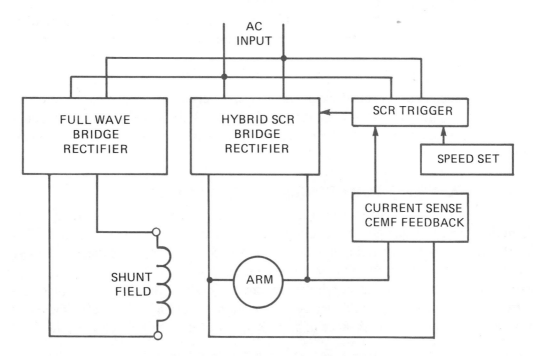

Fig. 3–3 Block diagram of electronic speed control for dc motor

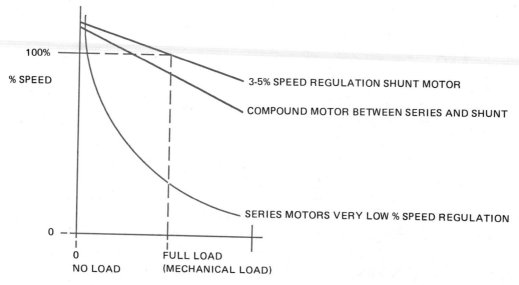

Fig. 3-4   Graph of speed regulation of shunt, compound, and series-connected dc motors

Speed Regulation

The speed regulation of a cumulative compound-wound motor is inferior to that of a shunt motor and superior to that of a series motor. It is a compromise between a series motor and a shunt motor, as can be seen in figure 3-4.

The graph in figure 3-4 shows that the percent of speed regulation of a compound wound dc motor is lower than that of a shunt motor but higher than that of a series motor.

ROTATION

The rotation of a compound-wound motor can be reversed by changing the direction of the current in the field or the armature circuit, (figure 3-5). Since the series field coils must also be reversed if the shunt field is reversed, it is easier to reverse the current in the armature only.

PRELIMINARY TEST FOR CUMULATIVE COMPOUNDING

When a motor is first connected, it is important to determine the continuity of the shunt field circuit. In addition, for a compound-wound motor, the proper magnetic polarity of the shunt and series field must be determined. Standardized tests determine these conditions. For example, when the motor is connected to the controller and is ready for starting, disconnect the armature wire at the motor, close the line switch, and place the starter on the first contact point. Open the line switch slowly. If the field is intact there will be an arc at the switch. The absence of a spark indicates an open field circuit. This fault

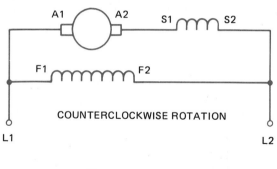

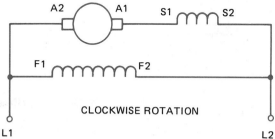

Fig. 3–5   Standard connections for compound motors (From Herman/Alerich, *Industrial Motor Control*, 2nd Ed., copyright 1990 by Delmar Publishers Inc.)

must be located and corrected before proceeding. A motor ordinarily will not start on an open field, but if it does start, it will race.

When the shunt field circuit tests complete, the motor should be started as a shunt motor. If the motor operates satisfactorily in the *desired* direction of rotation, disconnect the motor. If it rotates in the opposite direction, disconnect it and reverse the shunt field leads. Restart the motor. If it now rotates in the desired direction, mark the leads and disconnect power.

Next, open the shunt field circuit, connect in the series field, and operate the motor momentarily as a series machine. As soon as the armature begins to turn, note the direction of rotation and disconnect the motor power.

If the armature rotates in the desired direction, connect in the shunt field circuit and the motor is ready for operation. If the direction of rotation as a series motor is opposite to the desired direction, reverse the series field leads and then connect the shunt field circuit. The motor is now ready for operation. This makes sure the shunt field and series field help each other.

Differential Compounding

Excellent speed regulation can be obtained with a differential compound motor. When a motor is connected as a differential compound machine, the series field opposes the shunt field so that the field flux is decreased as a load is applied (figure 3–6). As a result, the speed remains substantially constant with an increase in load. With overcompounding, a slight increase in speed is possible with an increase in load. This speed characteristic is achieved only with a loss in the rate at which torque increases with load.

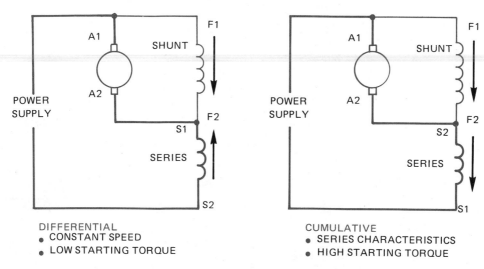

Fig. 3–6 **Magnetic polarities of compound motors**

Since the field decreases with a load increase, a differential compound motor has a tendency to speed instability. When starting a differential motor, it is recommended that the series field be shorted since the great starting current in this field may overbalance the shunt field and cause the motor to start in the opposite direction.

A differential machine is connected and tested on installation, using the same procedure outlined for a cumulative compound motor. For the differential motor, however, the series windings should be connected in the opposite direction from that of the shunt winding. Do not exceed load on nameplate or reversal of direction may occur.

SUMMARY

The dc compound motor is used where a compromise is needed between the series and the shunt motor. The compound motor has better speed characteristic than the series motor and better torque characteristics than the shunt motor. The motor can be connected so that the shunt field and the series fields are in the same direction. This is the *cumulative connection.* You should perform a test to be sure the motor is connected as intended. The cumulative connection reacts quite differently than the differential connection. Many dc motors are now driven by electronic drives. The same basic concepts are used to control speed and direction.

ACHIEVEMENT REVIEW

1. Circle the letter for each of the following statements which applies to a cumulative compound-wound dc motor.
    a. The speed regulation of a cumulative compound-wound dc motor is better than that of a shunt motor.

b. The speed of a cumulative motor has a no-load limit.

c. The speed of the motor decreases more for a given increase in load than does a differential motor.

d. A cumulative motor has less torque than a shunt motor of the same hp rating for a given increase in armature current.

e. The speed regulation of a cumulative motor is better than that of a series motor.

f. A cumulative motor develops a high torque with a sudden increase of load.

g. To reverse the direction of rotation, the current in either the armature or the shunt field must be reversed.

h. A cumulative motor is connected so that the series flux aids the shunt winding flux.

i. When installing a cumulative compound-wound motor, the direction of rotation should be the same when testing the motor for operation either as a series motor or a shunt motor.

2. Circle the letter for each of the following statements which applies to a differential compound-wound dc motor.

a. A differential motor is used in applications where an essentially constant speed at various loads is required.

b. The starting torque for a differential motor is higher than that of a cumulative motor.

c. The motor may reverse its direction of rotation if started under a heavy load.

d. This motor develops a speed instability since the flux field decreases with a load increase.

e. When starting a differential motor, the shunt field should be shorted because of the great starting current.

3. Arrange the following steps numerically in the correct sequence to test for the proper connections to operate a cumulative compound-wound motor. Place the step number in the space provided, starting with number 1.

_____ a. Place starter on first contact point to test field.

_____ b. With motor shut down, open shunt field circuit.

_____ c. If rotation is in direction opposite to that desired, reverse the series field leads.

_____ d. Disconnect armature wire at motor and close line switch.

_____ e. An absence of spark indicates an open field circuit which must be located and corrected before proceeding with the test.

_____ f. If rotation is in direction opposite to that desired, shut down motor, reverse shunt field leads, and restart motor; rotation should be in desired direction.

_____ g. Slowly open line switch and observe for arc at switch indicating field is intact.

_____ h. Start motor as a shunt motor and observe rotation.

_____ i. Connect in series field, start motor and immediately shut it down while noting direction of rotation.

_____ j. Connect the shunt field circuit; the motor is now ready for operation.

4

SUMMARY REVIEW
OF UNITS 1-3

OBJECTIVE

- To give the student an opportunity to evaluate the knowledge and understanding acquired in the study of the previous three units.

Select the correct answer for each of the following statements and place the corresponding letter in the space provided.

1. The twisting effect of a motor shaft is called its _____
   a. turning power.   c. r/min.
   b. horsepower.   d. torque.

2. The twisting effect of a dc motor is produced primarily by _____
   a. the armature.
   b. the rotor.
   c. a current-carrying conductor in a magnetic field.
   d. torque in the field coils.

3. A dc motor is required to maintain the same speed at no load and full load. This type of operation can only be obtained by using a _____
   a. series motor.   c. differential compound-wound motor.
   b. shunt motor.   d. cumulative compound-wound motor.

4. As a load is applied to a dc shunt motor the _____
   a. field current increases.
   b. counter emf increases.
   c. armature current increases.
   d. torque developed decreases.

5. The speed of a dc shunt motor _____
   a. increases with an increase in load.
   b. decreases with an increase in applied voltage.
   c. decreases if the field strength is increased.
   d. decreases less than a series motor of the same hp for the same increase in load.

6. As load is applied to a dc series motor the
    a. field current decreases.
    b. field voltage increases.
    c. armature current decreases.
    d. armature voltage increases.
    _____

7. The load requirements of a particular dc motor installation require extremely high starting torque. If speed regulation is not important, use a
    a. series motor.
    b. shunt motor.
    c. differential compound-wound motor.
    d. cumulative compound-wound motor.
    _____

8. As a load is applied to a cumulative compound-wound dc motor its
    a. speed decreases.
    b. counter emf decreases.
    c. torque decreases.
    d. series field current decreases.
    _____

9. In a cumulative compound-wound dc motor, the
    a. series winding develops the major part of the total flux.
    b. series and shunt windings develop field flux in the same direction.
    c. shunt winding must be connected across the brushes.
    d. series windings do not pass the shunt field current.
    _____

10. In a dc shunt motor *all but one* of the following are true.
    a. Torque is proportional to the field current.
    b. The same voltage is applied to armature and field circuits.
    c. The no-load speed is controlled by the impressed voltage.
    d. The motor is suitable for installations requiring substantially constant speed with variable loading.
    _____

11. In a differential compound-wound dc motor
    a. the series and shunt fields establish flux in the same direction.
    b. the series winding acts to reduce speed as load is applied.
    c. an increase in total current input as the result of loading increases the shunt field current.
    d. changes in torque result in change of current in the series field windings.
    _____

12. If the direction of field flux and the direction of armature
    current are changed, the torque developed by the motor is   _____
    a. stronger.   c. the same.
    b. less.   d. reversed.

13. A generator shunt field winding is   _____
    a. high resistance.   c. noninductively wound.
    b. low resistance.   d. embedded in the armature.

14. For proper operation in a four-lead dc motor, leads S_1 and
    S_2 should be connected to A_1 and A_2 in   _____
    a. parallel.   c. shunt.
    b. series.   d. series-parallel.

15. Decreasing the resistance of a generator field rheostat
    a. decreases the flux.   c. increases the voltage.
    b. decreases the voltage.   d. decreases the speed.

16. A series field, if connected across a motor armature and energized,   _____
    a. makes the motor race dangerously.
    b. causes a short circuit.
    c. creates excessive flux.
    d. acts as a shunt field.

U • N • I • T
5

MANUAL STARTING RHEOSTATS FOR DC MOTORS

OBJECTIVES

After studying this unit, the student will be able to

- list the primary functions of the three-terminal starting rheostat and the four-terminal starting rheostat.
- demonstrate the proper connection and startup procedure for a three-terminal starting rheostat on a shunt motor and on a cumulative compound-wound motor.
- demonstrate the proper connection and startup procedure for a four-terminal starting rheostat on a shunt motor and on a cumulative compound-wound motor.
- state basic dc motor starting principles common to other motor starters.

Manual starting rheostats for dc motors are becoming more and more rare. Electronic controls have replaced the manual three-point and four-point controllers for dc motor starters. However, three-point and four-point controllers are still being used, so the maintenance electrician should know the principles of operation of the manual dc starters.

Two factors limit the current taken by a motor armature from a direct-current source: (1) the counter electromotive force, and (2) the armature resistance. **Since there is no counter emf when the armature is at a standstill, the current taken by the armature will be abnormally high. As a result, the armature current must be limited by an external resistor, such as a starting rheostat.** *Electric Motor Control* covers in detail this method of limiting armature current.

A starting rheostat or motor starter is described by the National Electrical Manufacturers' Association as a device designed to accelerate a motor to its normal rated speed in one direction of rotation. In addition, a motor starter limits the current in the armature circuit to a safe value during the starting or accelerating period. Two types of manual starting rheostats are:

- three-terminal starting rheostat, and
- four-terminal starting rheostat.

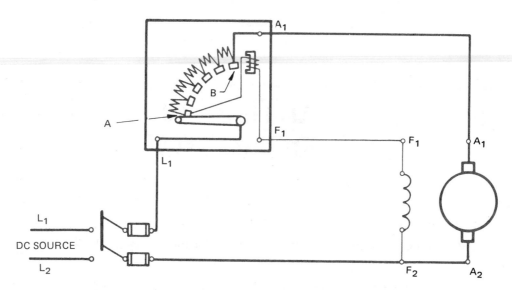

Fig. 5–1   Connections for a three-terminal starting rheostat

The electrical technician is expected to know how each type of starting rheostat is connected to direct-current shunt and compound motors. The technician should also know the specific applications and limitations of each type of motor starter.

THREE-TERMINAL STARTING RHEOSTAT

The three-terminal starting rheostat has a tapped resistor enclosed in a ventilated box. Contact buttons located on a panel mounted on the front of the box are connected to the tapped resistor. A movable arm with a spring reset can be moved over the contact buttons to cut out sections of the tapped resistor.

The connection diagram for a typical three-terminal starting rheostat is shown in figure 5–1. Note that the starter has three terminals or connection points and that it is connected to a shunt motor.

When the arm of the rheostat is moved to the first contact, A, the armature (which is in series with the starting resistance) is connected across the source. The shunt field, in series with the holding coil, is also connected across the source. The initial current inrush to the armature is limited to a safe value by the starting resistance. In addition, the shunt field current is at a maximum value and provides a good starting torque.

As the arm is moved to the right toward contact B, the starting resistance is reduced and the motor accelerates to its rated speed. When the arm reaches contact B, the armature is connected directly across the source voltage, and the motor will have attained full speed.

The holding coil is connected in series with the shunt field and provides a nofield release. **If the shunt field opens, the motor speed will become dangerously high if the armature circuit remains connected across the source.** Therefore, in the event of an open-circuited shunt field, the holding coil of the starting rheostat becomes demagnetized and the arm returns to the off position.

Note that the starting resistance is in series with the shunt field when the arm is in the run position at contact B. This additional resistance has practically no effect on the speed as the starting resistance is small when compared with the shunt field resistance.

To operate a three-terminal starter, first close the line switch. **Then, move the starting arm from the off position to contact A. Continue to move the arm slowly toward contact B, pausing on each intermediate contact for a period of one to two seconds.** By moving the arm slowly toward the run position at contact B, the motor will accelerate uniformly to its rated speed without an excessive inrush of current to the armature. However, do not hold the arm on any one contact between A and B for too long a period of time. These starting resistors are designed to carry the starting current for a short period of time only. In other words, do not control the speed of the motor by holding the arm for any length of time on any contact between A and B.

If it is necessary to control the speed of the motor, **do not use a three-terminal starter**. Since the current in the shunt field and holding coil may be reduced to a value insufficient to hold the arm against the action of the reset spring, the reset spring will return the arm to the off position. Thus, the motor will become disconnected from the current source.

When a motor is to be disconnected from the current source, first open the line switch quickly. Then, check to see that the spring reset returns the starting arm to the off position.

Figure 5–2 shows the connections for a three-terminal starting rheostat used with a cumulative compound-wound motor. Note that these connections are almost the same as

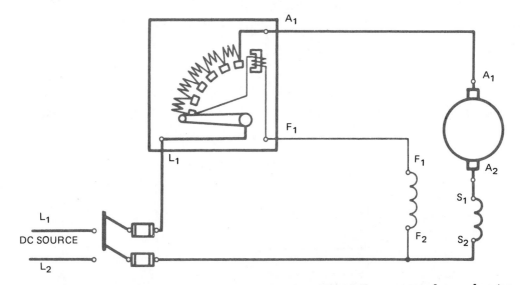

Fig. 5–2   Three-terminal starting rheostat connected to a cumulative compound-wound motor

those of a three-terminal starting rheostat connected to a shunt motor; the only change in figure 5–2 is the addition to the motor of the series field.

FOUR-TERMINAL STARTING RHEOSTAT

A four-terminal starting rheostat performs the same functions as a three-terminal starting rheostat:

- it accelerates a motor to rated speed in one direction of rotation.
- it limits the starting surge of current in the armature circuit to a safe value.

In addition, the four-terminal starting rheostat may be used where a wide range of motor speeds is necessary. A field rheostat may be inserted in series with the shunt field circuit to obtain the desired speed.

Figure 5–3 shows a four-terminal starting rheostat. Note that the holding coil is not connected in series with the shunt field as it is in the three-terminal starting rheostat. The holding coil in figure 5–3 is connected in series with a resistor across the source. Note also that the holding coil circuit is connected across the source and, as a result, four terminal connection points are necessary.

The holding coil of the rheostat is connected across the source and acts as a novoltage release. For example, if the line voltage drops below the desired value, the attraction of the holding coil is decreased, and the reset spring will then return the arm to the off position.

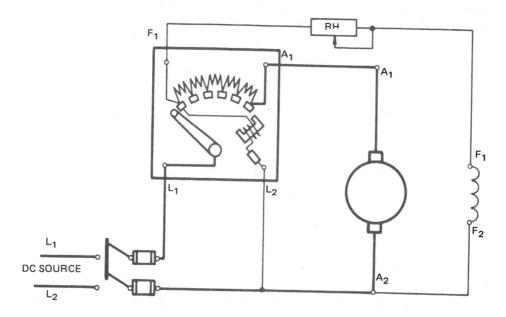

Fig. 5–3   Connections for a four-terminal starting rheostat

When a four-terminal starting rheostat is used, the speed of the motor is controlled by varying the resistance of the field rheostat connected in series with the shunt field circuit. The speed *is increased* above the rated speed by inserting resistance in the field rheostat.

When a motor using a four-terminal starting rheostat is to be disconnected from the source, **first cut out all resistance in the field rheostat. Then, open the line switch and check to see that the spring reset returns the starting arm to the off position.**

By removing all resistance from the field rheostat, the strength of the shunt field is increased. Thus, when the motor is restarted, it will have a strong field and strong starting torque.

Figure 5–4 shows a four-terminal starting rheostat connected to a cumulative compound-wound motor. Note the similarity in connections for a shunt motor (figure 5–3) and a compound-wound motor (figure 5–4). The only change in figure 5–4 is the addition of the series field.

NATIONAL ELECTRICAL CODE RULES FOR MOTOR STARTERS

The National Electrical Code states that a motor starter shall be marked with voltage and horsepower ratings, and the manufacturer's name and identification symbols, such as style or type numbers.

Electrical codes require that the horsepower rating of a starter must not be smaller than the horsepower rating of the motor. In addition, the fuse protection for dc motors must be no greater than a percentage of the full-load current rating of the motor. Therefore, the motor starter must limit the starting current to a value which is no greater than a percentage (specified by the electrical code) of the full-load current rating of the motor.

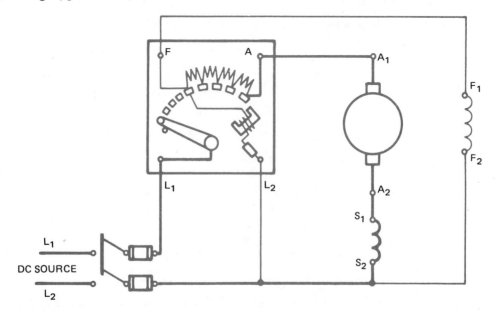

Fig. 5–4   Four-terminal starting rheostat connected to a cumulative compound-wound motor,

It is recommended at this time to review the **National Electrical Code** section on motors, motor circuits, and controllers.

SAFETY PRECAUTIONS FOR DC MOTOR STARTERS

Current must be limited when starting the armature. A no-field release is provided to prevent runaway acceleration.

To operate a three- or four-terminal starter, the starting arm is moved slowly, pausing for one or two seconds only before moving to the next position.

Three-Terminal Starter

To decelerate the motor, first open the line switch. Check the starting arm and return it to the OFF position by using the reset spring.

Four-Terminal Starting Rheostat

Decelerate the motor by cutting out all resistance in the field rheostat. Open the line switch. Check the starting arm and return it to the OFF position by using the reset spring.

SUMMARY

Three-terminal and four-terminal starting rheostats are not used too much any more. The advent of electronic starters have all but replaced the mechanical starters. The concept of the starter is still used and the safeguards for motor operation are still important.

ACHIEVEMENT REVIEW

1. What are the two functions of a motor starter?

    a. _____

    b. _____

2. Show the connections of a three-terminal starting rheostat to a shunt motor.

3. State one advantage of a three-terminal starting rheostat.

    _____

    _____

4. Name one limitation of a three-terminal starting rheostat.

   _____

   _____

5. Complete the connections in the following figure to show the shunt motor properly connected to the three-terminal starting rheostat.

THREE-TERMINAL
STARTING RHEOSTAT

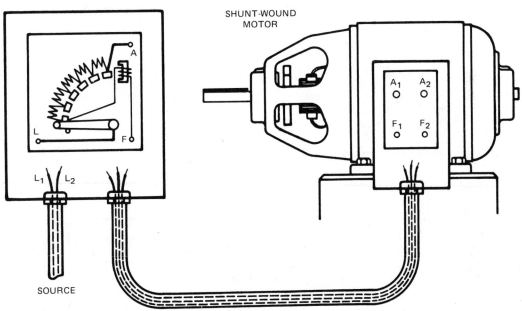

6. Show the connections of a four-terminal starting rheostat to a shunt motor.

7. What is one advantage of a four-terminal starting rheostat?

   _____

   _____

8. List the items that should be marked on the nameplate of a motor starter to comply with National Electrical Code requirements.

   _____

   _____

9. Complete the connections in the following figure to show that the cumulative com-
    pound-wound motor can be started from the four-terminal starting rheostat. Also
    connect the field rheostat in the circuit for above-normal speed control.

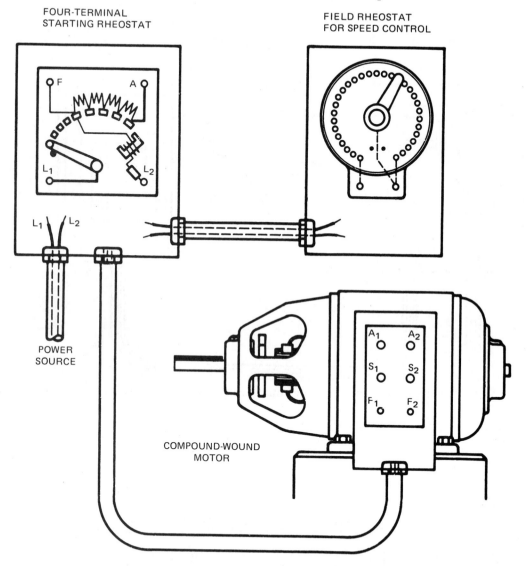

10. What is the full-load current rating of a 5-hp, 240-V motor? (Refer to the National
    Electrical Code, if necessary.)

    _____

11. What size conduit is required between the 5-hp motor and the starting box, using
    type THHN wire?

    _____

U • N • I • T
6

SPECIAL DC STARTING RHEOSTATS AND CONTROLLERS

OBJECTIVES

After studying this unit, the student will be able to

- describe the operation of a series motor starter with no-voltage protection.
- describe the operation of a series motor starter with no-load protection.
- describe the actions occurring at each forward and reverse position of a drum controller.

Series motors require a special type of starting rheostat called a *series motor starter*. These starting rheostats serve the same purpose as the three- and four-terminal starting rheostats used with shunt and compound motors. However, the internal and external connections for the series motor starter differ from the connections of the other types of starting rheostats.

Series and cumulative compound motors are often used for special industrial applications which require provisions for reversing the direction of rotation and varying the speed of the motor. A manually operated controller, called a *drum controller*, may be used for these applications. The operation of drum controllers is discussed later in this unit.

STARTING RHEOSTATS FOR DC SERIES MOTORS

Series motor starting rheostats are of two types: one type of starter has no-voltage protection, and the other type has no-load protection.

Starter with No-Voltage Protection

A series motor starter with no-voltage protection is shown in figure 6–1. The holding coil circuit of this starter is connected across the source voltage. There is no shunt field connection on this type of starter as it is used only with series motors. This type of starter is used to accelerate the motor to rated speed. In the event of voltage failure, the holding coil no longer acts as an electromagnet. The spring reset then quickly returns the arm to the off position. Thus, the motor is protected from possible damage due to low-voltage conditions.

To disconnect a motor using this type of starting rheostat, open the line switch. Check to be sure that the arm returns to the off position.

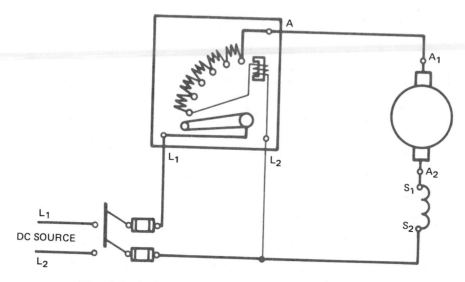

Fig. 6–1   Series motor starter with no-voltage protection

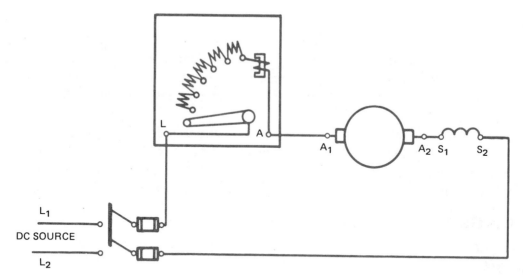

Fig. 6–2   Series motor starter with no-load protection

Starter with No-Load Protection

A series motor starter with no-load protection is shown in figure 6–2. The holding coil is in series with the armature circuit. Because of the relatively large current in the armature circuit, the holding coil consists of only a few turns of heavy wire. Note in the figure that separate terminal connections for the shunt field and holding coil are not provided. There are only two terminals–one marked L (line) and one marked A (armature).

The same care is required in starting a motor with this type of starting rheostat as is required with three- and four-terminal starting rheostats. The arm is slowly moved from the off position to the run position, pausing on each contact button for a period of one to two seconds. The arm is held against the tension of the reset spring by the holding coil connected in series with the armature. If the load current to the motor drops to a low value, the holding coil weakens and the reset spring returns the arm to the off position. This is an important protective feature. **Recall that a series motor may reach a dangerously high speed at light loads.** Therefore, if the motor current drops to such a low value that the speed becomes dangerous, the holding coil will release the arm to the off position. In this way, it is possible to avoid damage to the motor due to excessive speeds.

To stop a series motor connected to this type of starting rheostat, open the line switch. Check to be sure that the arm returns to the off position.

DRUM CONTROLLERS

Drum controllers are used when an operator is controlling the motor directly. The drum controller is used to start, stop, reverse, and vary the speed of a motor. This type of controller is used on crane motors, elevators, machine tools, and other applications in heavy industry. As a result, the drum controller must be more rugged than the starting rheostat.

A drum controller with its cover removed is shown in figure 6–3. The switch consists of a series of contacts mounted on a movable cylinder. The contacts, which are insulated

Fig. 6–3   Drum type controller shows contact fingers

from the cylinder and from one another, are called movable contacts. There is another set of contacts, called stationary contacts, located inside the controller. These stationary contacts are arranged to touch the movable contacts as the cylinder is rotated. A handle, keyed to the shaft for the movable cylinder and contacts, is located on top of the drum controller. This handle can be moved either clockwise or counterclockwise to give a range of speed control in either direction of rotation. The handle can remain stationary in either the forward or reverse direction due to a roller and a notched wheel. A spring forces the roller into one of the notches at each successive position of the controller handle to keep the cylinder and movable contacts stationary until the handle is moved by the operator.

A drum controller with two steps of resistance is shown in figure 6–4. The contacts are represented in a flat position in this schematic diagram to make it easier to trace the circuit connections. To operate the motor in the forward direction, the set of contacts on

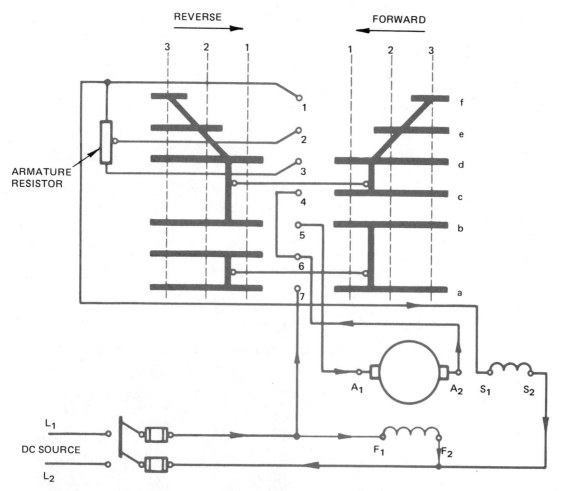

Fig. 6–4   Schematic diagram of a drum controller connected to a compound-wound motor

the right must make contact with the center stationary contacts. Operation in the reverse direction requires that the set of movable contacts on the left makes contact with the center stationary contacts.

Note in figure 6–4 that there are three forward positions and three reverse positions to which the controller handle can be set. In the first forward position, all of the resistance is in series with the armature. The circuit path for the first forward position is as follows:

1. Movable fingers a, b, c, and d contact the stationary contacts 7, 5, 4, and 3.

2. The current path is from the positive side of the line to contact 7, from 7 to a, from a to b, from b to 5, and then to armature terminal A_1.

3. After passing through the armature winding to terminal A_2 the current path is to stationary contact 6, and then to stationary contact 4.

4. From contact 4 the current path is to contact c, to d, and then to contact 3.

5. The current path then goes through the armature resistor, to the series field, and then back to the negative side of the line.

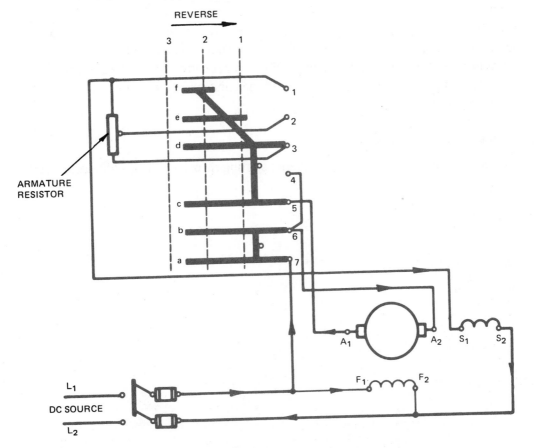

Fig. 6–5   First position of controller for reverse direction

The shunt field of the compound motor is connected across the source voltage. On the second forward position of the controller handle, part of the resistance is cut out. The third forward position cuts out all of the resistance and puts the armature circuit directly across the source voltage.

In the first reverse position, all of the resistance is inserted in series with the armature.

Figure 6–5 shows the first position of the controller in the reverse direction. The current in the armature circuit is reversed. However, the current direction in the shunt and series fields is the same as the direction for the forward positions. Remember that an earlier unit showed that a change in current direction in the armature *only* resulted in a change in the direction of rotation.

The second reverse position cuts out part of the resistance circuit. The third reverse position cuts out all of the resistance and puts the armature circuit directly across the source. Drum controllers with more positions for a greater control of speed can be obtained. However, these controllers all use the same type of circuit arrangement shown in this unit.

SUMMARY

DC series motors require a different starting controller than shunt or compound motor. The holding circuit for the controller is in series with the starting resistance. If there is a low-voltage or no-voltage condition, the starter is returned to the off position. Drum controllers are still used frequently. Often drum controllers are used with ac as well as dc motors. It is important to be able to read the connection diagrams and the sequence diagrams on drum-type controllers.

ACHIEVEMENT REVIEW

1. Show the internal connections of a series starting rheostat with no-voltage protection.

2. Show the internal connections of a series starting rheostat with no-load protection.

3. Show the circuit connections for a series motor used with a series starting rheostat with no-load protection.

4. Show the circuit connections for a series motor used with a series starting rheostat with no voltage protection.

5. Why is a drum controller used in many industrial applications?

   _____

   _____

   Complete the following statements.

6. In a series starter with no-voltage protection, the holding coil is connected across the

   _____.

7. A series starter with no-load protection is used to prevent the series motor from reaching _____ at low loads.

8. A drum controller gives the following types of control for a direct-current motor:

_____

_____.

BASIC PRINCIPLES OF AUTOMATIC MOTOR CONTROL

OBJECTIVES

After studying this unit, the student will be able to

- list several factors to be considered when selecting and installing electric motor control equipment.

- explain the purpose of a contactor.

- describe the basic operation of a contactor and relay.

- list the steps in the operation of a control circuit using start and stop pushbuttons.

- interpret simple automatic control diagrams.

- draw a simple magnetic control circuit.

Motor control was a simple problem when motors were used to drive a common line shaft to which several machines were connected. In this arrangement, it was necessary to start and stop only a few times daily.

With individual drive, however, the motor is an integral part of the machine and the motor controller must be designed to meet the needs of the machine to which it is connected.

As a result, the modern motor controller does not just start, stop, and control the speed of a motor. The controller may also be required to sense a number of conditions, including changes in temperature, open circuits, current limitations, overload, smoke density, level of liquids, or the position of devices. Manual control is limited to pressing a button to start or stop the entire sequence of operations at the machine or from a remote position.

The electrician must know the symbols and terms used in automatic control diagrams to be able to wire, install, troubleshoot, and maintain automatic control equipment.

CLASSIFICATION OF AUTOMATIC CONTROLLERS
Purpose

Factors to be considered in selecting motor controllers include what types of starting, stopping, reversing, running, speed and sequence control, and protection are required.

Fig. 7–1A   DC magnetic relay

Fig. 7–1B   DC operated relay (*Courtesy of Square D Company*)

Operation

The motor may be controlled directly or manually by an operator using a switch or a drum controller. Remote control uses contactors, relays, and pushbuttons, sensors and possibly electronics.

CONTACTORS

Contactors, or relays (figure 7–1) are required in automatic controls to transmit varying conditions in one circuit to influence the operation of other devices in the same or another electrical circuit. Relays have been designed to respond to one or more of the following conditions:

voltage	overvoltage	undervoltage
current	overcurrent	undercurrent
current direction	differential current	power (watts)
power direction	volt-amperes	frequency
phase angle	power factor	phase rotation
phase failure	impedance	speed
	temperature	

Magnetic switches are widely used in controllers because they can be used with remote control, and are economical and safe.

A relay or contactor usually has a coil which can be energized to close or open contacts in an electrical circuit. The coil and contacts of a relay are represented by symbols on the circuit diagram or schematic of a controller. Symbols commonly used to represent contactor elements are shown in figure 7–2.

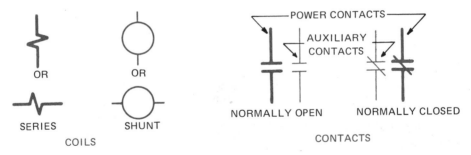

Fig. 7–2   **Schematic symbols for contactor elements**

If the control coil is connected in series in the motor power circuit, the heavy line symbol shown at the left of figure 7–2 is used. If the coil is connected in parallel (shunt), the light line symbol is used.

A series coil has a large current carrying conductor with few turns designed to carry large currents. A shunt coil has a small wire size with many turns; it carries small currents. It is possible for a series coil and a shunt coil to have the same ampereturns, resulting in similar magnetic results.

Contacts which are open when the coil is deenergized are known as *normally open* contacts and are indicated by two short parallel lines. Contacts which are closed when the coil is deenergized are called *normally closed* contacts and are indicated by a slant line drawn across the parallel lines.

To minimize heavy arcing which burns the contacts, a dc contactor usually is equipped with a *blowout coil* and an *arc chute*. Figure 7–3 shows a magnetic contactor which is provided with a blowout coil and an arc chute.

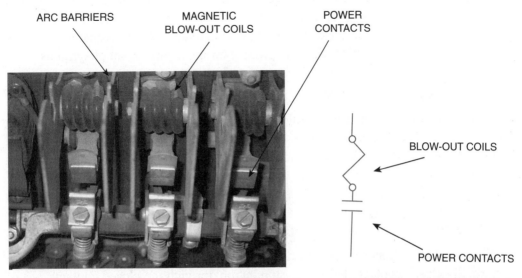

Fig. 7–3   **Magnetic blow out coils magnetically move the arc away from the contacts (From Keljik, *Electric Motors and Motor Controls*, copyright 1995 by Delmar Publishers)**

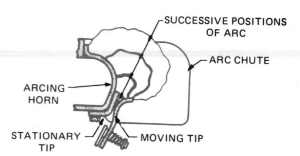

Fig. 7–4 Beharvior of arc with correctly designed blowout

When a heavy current is broken by the contacts of the contactor, an arc occurs. Figure 7–4 illustrates the behavior of the arc as it is quickly extinguished by the electromagnetic and thermal action of the magnetic blowout coil and arc chute.

PUSHBUTTONS

Pushbutton stations (figures 7–5A and B), are spring-controlled switches and, when pushed, are used to complete motor or motor control circuits. Figure 7–5B shows multiple control stations, with pushbuttons, selector switches, and pilot indicating lights. Note the "mushroom" stop button for easy access. This is for convenience and safety.

The symbols used in schematic, drawings to represent momentary pushbutton contacts are given in figure 7–6. Contacts can either be normally-open or normally-closed. This is the normal condition when there is no mechanical actuation of the contacts. In the pushbuttons shown in figure 7–6, the contacts are referred to as *momentary contacts*. This simply means that the contacts change from their *normal* condition to the opposite condition momentarily when mechanical actuation is applied, and then change back to the normal condition when the actuator is removed. Some contacts are designated as *maintained* contacts. This means that the contacts will stay as activated (held mechanically) until returned to their original position. See the glossary for the complete set of symbols.

TYPICAL CONTROL CIRCUIT

Figure 7–7 is an elementary control circuit with start and stop buttons and a sealing circuit. The following sequence describes the operation of the circuit. The typical control circuit uses an electromagnetic coil to move sets of contacts. The contacts move to open and close the power circuit to the motor, and also open and close contacts in the control circuit. The control contacts provide a sealing circuit in parallel to the start momentary contacts. This parallel circuit is referred to as the *sealing circuit*. It seals a current path around the normally open start button contacts. The circuit operation is as follows:

1. When the start button is pressed to close contacts 2–3, current flows from L_1 through normally closed contacts 1–2 of the stop button, through the closed contacts 2–3 of the start button, and through coil M to Line L_2.

Fig. 7–5   Pushbutton stations (*Courtesy of Square D Company*)

NORMALLY OPEN

NORMALLY CLOSED

OPEN AND CLOSED

Fig. 7–6   Symbols for pushbutton contacts

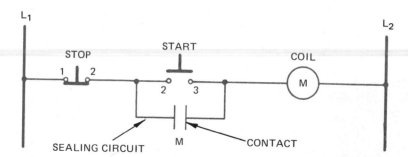

Fig. 7–7 A control circuit with start and stop buttons and sealing circuit

2. The current in coil M causes the contact M to close. Thus, the sealing circuit around contacts 2–3 of the start button closes. The start button may now be released, and even though the spring of the pushbutton opens contacts 2–3, coil M remains energized and holds contacts M closed to maintain a sealing circuit around the normally open contacts 2–3 of the start button. Coil M, being energized, also closes M contacts in the power circuit to the motor (not shown).

3. If the stop button is momentarily pressed, the circuit is interrupted at contacts 1–2 and coil M is deenergized. Contacts M then open and coil M cannot be energized until the start button again closes contacts 2–3.

SUMMARY

The basic automatic control circuit is used to control larger motors and to control them through electromagnetic relays. This allows the operation to be remotely located and the contactor to be located near the motor. The basic principle uses a momentary-contact switch to close a circuit to a magnetically-operated relay.

ACHIEVEMENT REVIEW

Select the correct answer for each of the following statements and place the corresponding letter in the space provided.

1. Early motor installations consisted of   _____
    a. individual drives.
    b. a common line shaft drive.
    c. automatically controlled motor drives.
    d. remotely controlled motors.

2. Individual motor drives require   _____
    a. single-phase motors.
    b. automatic controllers.
    c. speed rheostats.
    d. gear heads.

3. Automatic dc motor controllers are designed to respond to changes in temperature, open circuits, current limitations, and _____
   a. wire size.   c. speed acceleration.
   b. fuse rating.   d. brush assembly.

4. Interpretation of automatic control circuits requires the recognition of _____
   a. color.   c. ratings
   b. electrical circuit symbols.   d. parallel circuits

5. A relay symbol shows the _____
   a. number of turns in a coil.
   b. relay current rating.
   c. relative position of the component parts.
   d. size of the contacts.

6. A relay is classified as a piece of electrical equipment with at least one _____
   a. coil.
   b. resistor.
   c. coil operating one contact.
   d. coil operating two contacts.

7. Normally open contacts are _____
   a. open at all times.
   b. open when the relay coil is deenergized.
   c. open when the relay coil is energized.
   d. contacts that open a circuit.

8. Normally closed relay contacts are represented by the symbol: _____

   a.   [symbol]   c.   [symbol]

   b.   [symbol]   d.   [symbol]

9. A sealing circuit bypasses _____
   a. the armature circuit.
   b. the field circuit.
   c. the ON pushbutton contacts.
   d. the relay coil.

10. Elementary control diagrams are read from _____
    a. top to bottom.   c. right to left.
    b. bottom to top.   d. field to armature circuit.

U • N • I • T

8

THE DC COUNTER EMF MOTOR CONTROLLER AND DC VARIABLE SPEED MOTOR DRIVES

OBJECTIVES

After studying this unit, the student will be able to

- explain the operation of the counter emf method of acceleration for a direct-current motor.
- make use of elementary wiring diagrams, panel wiring diagrams, and external wiring diagrams.
- explain the ratings of starting and running protection devices.
- describe the operating principles of dc variable speed motor drives.
- state how above and below dc motor speeds may be obtained.
- list the advantages of dc variable speed motor drives.
- describe how solid-state devices may replace rheostats.
- make simple drawings of dc motor drives.
- list the advantages of using thyristors.

Although manual starters are still used, most industrial applications use automatic motor control equipment to minimize the possibility of errors in human judgment. To install and maintain automatic motor control equipment the electrician must be familiar with three kinds of electrical circuit diagrams:

- elementary wiring diagram
- panel wiring diagram
- external wiring diagram

The elementary wiring diagram uses symbols and a simple plan of connections to illustrate the scheme of control and the sequence of operations.

The panel wiring diagram shows the electrical connections throughout all parts of the controller panel and indicates the external connections. All of the control elements are represented by symbols but are located in the same relative positions on the wiring diagram that they actually occupy on the control panel. Because of the maze of wires shown on the panel wiring diagram, it is difficult to use for troubleshooting or to obtain an understanding of the operation of the controller. For this reason, the elementary wiring diagram presents the sequence of operations of the controller and the panel diagram is used to locate problems and failures in the operation of the controller.

The external wiring diagram shows the wiring from the control panel to the motor and to the pushbutton stations. This diagram is most useful to the worker who installs the conduit and the wires between the starter panel and the control panel and motor.

COUNTER ELECTROMOTIVE FORCE METHOD OF MOTOR ACCELERATION CONTROL

The counter emf across the armature is low at the instant a motor starts. As the motor accelerates, this counter emf increases. The voltage across the motor armature can be used to activate relays which reduce the starting resistance when the proper motor speed is reached.

Starting and Running Protection for a Counter Emf Controller

Starting protection for a counter emf controller is provided by fuses in the motor feeder and branch-circuit line of the motor circuit. These fuses are rated according to *Article 430* of the NEC.

Running protection for a counter emf controller is provided by an overload thermal element connected in series with the armature. The thermal element is rated at 115 to 125 percent of the full-load armature current. As covered in NEC *Sections 430–32 and 430–34,* if the current exceeds the percent of the rated armature current value, heat produced in the thermal element causes the bimetallic strip to open or trip the thermal contacts which are connected in the control circuit. The value of current during the motor startup period does not last long enough to heat the thermal element sufficiently to cause it to open.

COUNTER EMF MOTOR CONTROL CIRCUIT

Starting the Motor (Refer to Figure 8–1A)

Close the main line switch before pressing the start button. After the start button is pressed, control relay M becomes energized. The control circuit is now complete from L_1 through the thermal overload (OL) contacts 6–7, through the start button contacts 7–8, through the normally closed stop button contacts 8–9 to L_2. The lower auxiliary sealing contacts 7–8 of relay M also close and bypass the start button. As a result, the start button may be released without disturbing the operation.

When the main contacts 2–3 of contactor M are closed, the motor armature circuit is complete from L_1 through overload thermal element contacts 1–2, through contacts 2–3 of relay M, through the starting resistor, and through armature leads 4–5 to L_2. The shunt field circuit $F_1 - F_2$ is connected in parallel with the armature circuit. Contacts 3–4 of counter emf contactor A remain open at startup because a high inrush current establishes a high voltage drop across resistor 3–4. This leaves only a small voltage drop across the armature and "A" contactor coil until acceleration is achieved.

Connecting the Motor across the Line

The counter emf generated in the armature is directly proportional to the speed of the motor. As the motor accelerates, the speed approaches the normal full speed and the counter emf increases to a maximum value. Relay A is calibrated to operate at approximately 80 percent of the rated voltage, When contacts 3–4 of relay A close, the starting resistance 3–4 is bypassed and the armature is connected across the line.

Running Overload Protection

A thermal overload relay contains two circuits. One circuit is in series with the armature and has the armature current flowing through its thermal sensor or heating element. The second circuit of the overload relay is the control circuit with a control contact. If the contact opens, because of excessive heat in the thermal heater, the control circuit will be interrupted, and stop the motor. A thermal overload relay unit is shown in figure 8–1B. The schematic diagram is shown in figure 8–1A.

When the load current of the armature exceeds the rated allowable percent of the full load current, the overload thermal element (points 1–2) heats up and opens contacts 6–7 in the control circuit. Control relay M is deenergized and main contacts 2–3 of M open and disconnect the motor from the line.

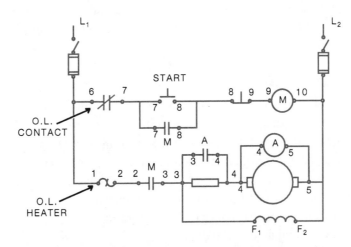

Fig. 8–1A Elementary diagram of a dc counter emf controller

HEATER:
BIMETAL DISK

HEATER:
MELTING ALLOY

HEATER:
BIMETAL STRIP

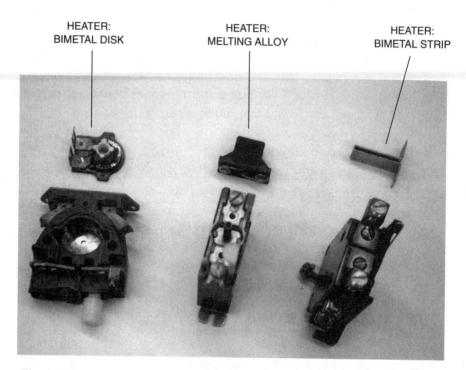

Fig. 8–1B   Bi-metal disk flexes as it is heated and opens motor control circuit. Photo of bimetal disk, solderpot type melting alloy, bi-metal strip (From Keljik, *Electric Motors and Motor Controls*, copyright 1995 by Delmar Publishers)

Stopping the Motor

When the stop button is pressed the control circuit is broken at points 8-9. The same shutdown sequence occurs as in the case of the overload condition discussed previously. The sealing circuit 7–8 is broken in each case.

The advantage of this type of automatic starter is that it does not supply full voltage across the armature until the speed of the motor is correct. The starter eliminates human error which may result from the use of a manual starter.

PANEL WIRING DIAGRAM

Figure 8–2 shows the same counter emf control circuit presented in figure 8-1A However, the panel wiring diagram locates the wiring on the panel in relationship to the actual location of the equipment terminals on the rear of the control panel. Troubleshooting or checking of original installations requires an accurate comparison of the elementary and panel diagrams. It is recommended that the electrician use a system of checking connections on the diagram with the actual panel connections. For example, a colored pencil may be used to make check marks on the diagram as each connection is properly traced on the panel and compared to the diagram.

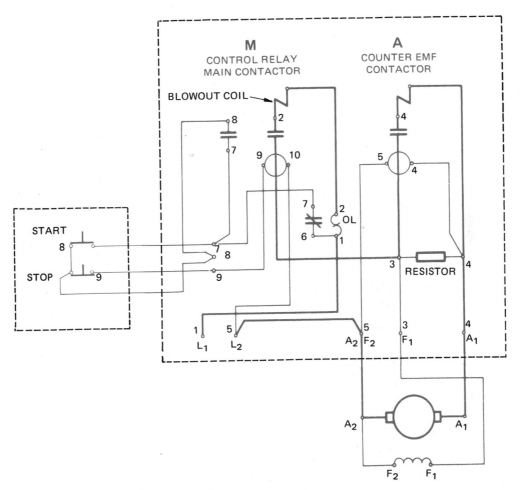

Fig. 8–2 Panel wiring diagram for a dc counter emf motor controller

CONDUIT OR EXTERNAL WIRING PLAN

All necessary external wiring between isolated panels and equipment is shown in the conduit plan (figure 8–3). The proper size of conduit, size and number of wires, and destination of each wire is indicated on this plan. An electrician refers to this plan when completing the actual installation of the counter emf controller.

DC ADJUSTABLE SPEED DRIVES

Dc adjustable speed drives are available in convenient units that include all necessary control and power circuits.

Some machinery requirements are so precise that the ac variable frequency drives may not be suitable (See unit 16). In such cases dc motors provide characteristics that are not available on ac motors. A dc motor with adjustable voltage control is very versatile and can be adapted to a large variety of applications.

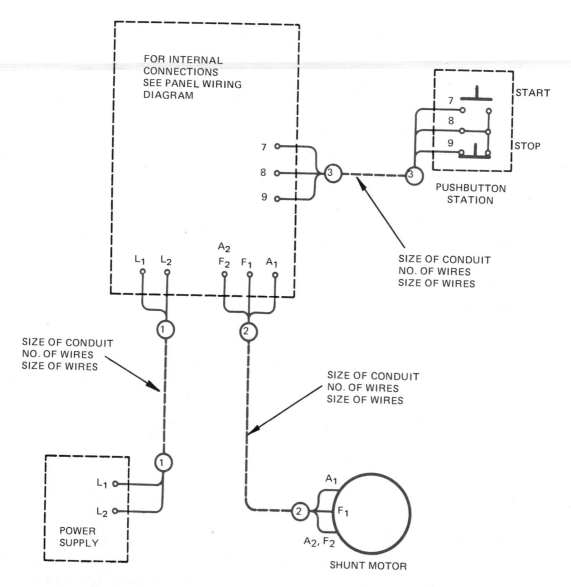

Fig. 8–3 Conduit or external wiring plan for a counter emf controller

In the larger horsepower range, the motor-generator set used to be one of the most widely used methods of obtaining variable speed control. The set consists of an ac motor driving a dc generator to supply power to a dc motor. Such motor-generator set drives, called Ward-Leonard Systems, control the speed of the motor by adjusting the power supplied to the field of the generator, and, as a result, the output voltage to the motor (figure 8–4). The generator field current can be varied with rheostats, as shown, or by variable transformers supplying a dc rectifier, or automatically with the use of solid-state controls. When it is desirable to control the motor field as well, similar means are used.

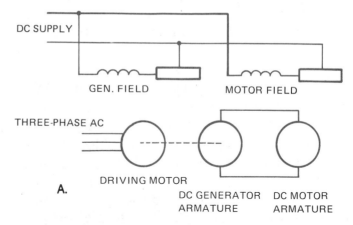

DC SUPPLY

GEN. FIELD   MOTOR FIELD

THREE-PHASE AC

A.

DRIVING MOTOR

DC GENERATOR   DC MOTOR
ARMATURE   ARMATURE

Fig. 8–4   A) Basic electrical theory of a dc motor-generator variable speed control system. B) Packaged motor-generator with dc variable speed control system supplied from ac (*Photo courtesy Square D Company*) (From Alerich, *Electric Motor Control*, 4th Ed., copyright 1988 by Delmar Publishers Inc.)

B.

The speed and torque of the system shown in figure 8–4 can be controlled by adjusting the voltage to the field, or to the armature, or both. Speeds *above* the motor base speed (nameplate speed) are obtained by weakening the motor shunt field. Speeds *below* the motor base speed are obtained by weakening the generator field. As a result, there is a decrease in the generator voltage supplying the dc motor armature. The motor should have a full shunt field for speeds lower than the base speed to give the effect of continuous control, rather than step control of the motor speed.

The motor used to furnish the driving power may be a three-phase induction motor, as shown in figure 8–4. After the driving motor is started, it runs continuously at a constant speed to drive the dc generator.

The armature of the generator is coupled electrically to the motor armature as shown. If the field strength of the generator is varied, the voltage from the dc generator can be controlled to send any amount of current to the dc motor. As a result, the motor can be made to turn at many different speeds. Because of the inductance of the dc fields and the time required by the generator to build up voltage, extremely smooth acceleration is obtained from zero r/min to speeds greater than the base speed.

The field of the dc generator can be reversed automatically, or manually, with a resulting reversal of the motor rotation.

The generator field resistance can be changed automatically by the use of SCRs (or thyristors) or time-delay relays operated by a counter EMF across the motor armature. The generator field resistance can also be changed manually.

Electrically controlled variable speed motor drives offer a wide choice of speed ranges, torque, and horsepower characteristics. They provide a means for controlling acceleration and deceleration, and methods of automatic or manual operation. A controlling tachometer feedback signal may be driven by the dc motor shaft. This is a system refinement to obtain a preset constant speed. This method depends upon the type of application, the speed, and the degree of response desired. In addition to speed, the controlling feedback signal may be set to respond to pressure, tension, shock, or some other transducer function.

One of the most advantageous characteristics of the motor-generator set drive is its inherent ability to regenerate. In other words, when a high inertia load overdrives the motor, the dc motor becomes a generator and delivers reverse power. For example, assume the dc motor is running at base speed. If the generator voltage is decreased by adjusting the rheostat to slow the motor, the motor counter voltage will be higher than the generator voltage and the current reverses. This action results in reverse torque in the motor and the motor slows down. This process is called dynamic braking. This dynamic feature is very desirable when used on hoists for lowering heavy loads, metal working machines, textile and paper processing machines, and for general industry for the controlled stopping of high inertia loads. Multiple motor drives are also accomplished with this type of motor-generator drive.

Motor-generator set drives using automatic regulators have been used for years for nearly every type of application. A higher degree of sophistication in controls has been developed, making it possible to meet almost any desired level of precision or response.

STATIC MOTOR CONTROL DRIVES

Despite the widespread acceptance and use of the motor-generator drives, rotating machines were required to convert ac to mechanical power. As a result, the combined efficiency of the set is rather low; it requires the usual rotating machine maintenance, and it is noisy. *Static* dc drives now being used have no moving parts in the power conversion equipment that converts (rectifies) and controls the ac power (figures 8–5 and 8–6). The solid-state devices are used for controlled conversion of ac line power to dc.

Fig. 8–5 Control panel for SCR controlled DC motor drive

Fig. 8–6 Silicon Controlled Rectifier (SCR) of various sizes

The basic theory for obtaining dc motor speeds below and above base speed are the same as with a motor-generator set. It is only the method of controlling the voltages and field strengths that differs. For example, in Figure 8–7, the armature is supplied with dc rectified from an ac source. The ac is rectified by the use of the S.C.R. in the controlled circuit to obtain dc. The gate of the thyristor will turn on the S.C.R. at the proper portion of the half wave, thereby controlling the motor below base speed. Figure 8–7 is a simplified circuit for the purpose of illustration. The field strength would be held at its fullest strength in a similar manner. For above motor base speed, the field control can weaken field strength with full armature voltage. The feedback tachometer will maintain a preset speed.

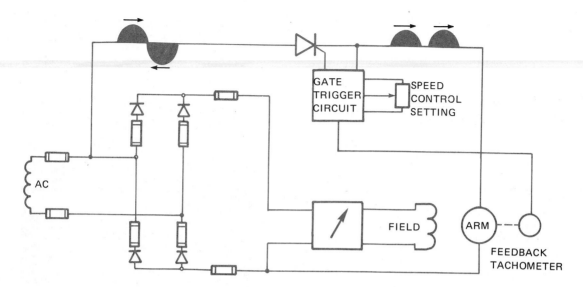

Fig. 8–7   Single-phase, half-wave armature controlling a small motor

In figure 8–7, the SCR is controlled by the setting of the potentiometer, *speed control*. This varies the "on" time of the thyristor per ac cycle, and thus varies the amount of average current flow to the armature. When speed control above the base speed is required, the rectifier circuit in the field is controlled by SCRs, rather than diodes.

The SCR, or thyristor, can control all of the positive waveform or voltage through the use of a method called *phase shifting*. It is not the intent of this text to cover the theory of the method.

The SCR is probably the most popular solid-state device for controlling large and small electrical power loads. The SCR is a controlled rectifier which controls an electric current. It will not conduct when the voltage across it is in the wrong direction. It will conduct only in the forward direction when the proper signal (voltage) is applied to the gate terminal. The gate is normally controlled by electronic pulses from a control circuit.

The gate will turn the SCR on but will not turn it off in a dc circuit. To turn the anode-cathode section of the SCR on (close the switch), the gate must be the same polarity as the anode with respect to the cathode. Once the gate has turned the SCR on, it will remain on until the current flowing through the power circuit (anode-cathode section) is either interrupted or drops to a low enough level to permit the device to turn off. The anode to cathode current must fall below the holding current level. The *holding current, or maintaining current,* is the amount of current required to keep the SCR turned on. The SCR performs the same function as a rheostat would in controlling motor field strengths or voltage to an armature. It is similar to a variable resistance, since it can be adjusted throughout its power range. The SCR control has replaced the rheostat since it is smaller in size for the same current rating, is more energy efficient, and is cheaper.

Figure 8–8 illustrates two sophisticated, single-phase, packaged, static control ac motor drive controls. They are shown with their wiring exposed and in their boxes. More elaborate units are available for three-phase ac power supplies.

SUMMARY

DC motors need controls to start, stop, protect, and adjust the speed and torque of the motor. The systems used must comply with the NEC and also have approval from testing

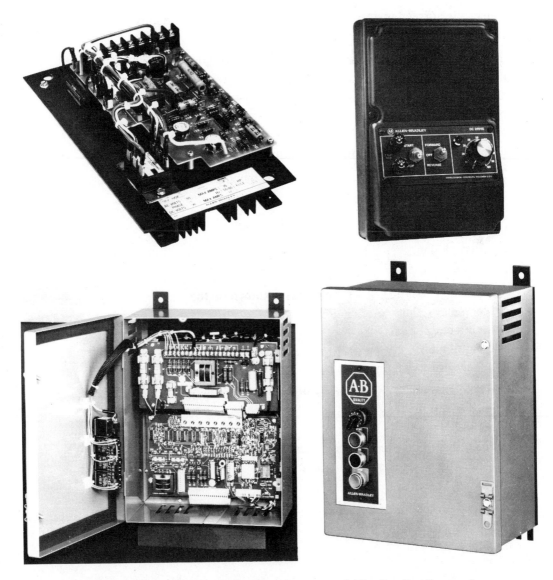

Fig. 8–8   Ac motor drive control (*Photos courtesy of Allen-Bradley Company*)

firms such as UL. The motors must be protected from overheating and causing damage to the motor and the surrounding area. This unit introduced the two general styles of wiring diagrams: the *schematic*, which shows the electrical location of the components, and the *wiring diagram* which shows the physical relationship of the equipment. Motor generator sets were presented to familiarize you with the possible sources of dc control. Now much of the control is done by solid state dc electronic drives.

ACHIEVEMENT REVIEW

Select the correct answer for each of the following statements and place the corresponding letter in the space provided.

1. The least important plan or diagram in troubleshooting motor
    controls is probably the
    a. elementary plan.
    b. panel diagram.
    c. external conduit plan.
    d. layout of the area in which the controllers are installed.

2. The best diagram to use to determine how a controller
    operates is the
    a. elementary plan.
    b. panel plan.
    c. external plan.
    d. architectural plan.

3. The physical location of control wires is shown on the
    a. elementary plan.
    b. architectural plan.
    c. conduit plan.
    d. panel wiring diagram.

4. The dc counter emf controller results from the automatic
    actions of the
    a. applied voltage.
    b. changing voltage across the armature.
    c. changing voltage across the field.
    d. starting current.

5. Overload protection is the same as
    a. starting protection.   c. electrical protection.
    b. mechanical protection.   d. running protection.

6. Overload contacts open the circuit when the motor
    current reaches   _____
    a. 85 percent of full load.
    b. 100 percent of full load.
    c. 125 percent of full load.
    d. 150 percent of full load.

7. In the event a motor is allowed to exceed the permissible
    current value, it is protected by   _____
    a. starting protection.
    b. fuses.
    c. an overload thermal element.
    d. the stop button.

8. With the disconnect switch closed, the shunt field in figure
    8–1 is placed across the line when the   _____
    a. A contact closes.
    b. disconnect switch is closed.
    c. M contact closes.
    d. start button closes.

9. In figure 8–1, contact A is closed when the   _____
    a. start button is closed.
    b. stop button is opened.
    c. A coil is deenergized.
    d. A coil is energized.

10. The motor in figure 8–1 is placed across the line when   _____
    a. the start button is closed.
    b. the disconnect switch is closed.
    c. contact A is closed.
    d. contact M is closed.

11. What is the dc motor base speed? _____
    _____

12. How is the speed of a dc motor controlled *above* the base speed? _____
    _____
    _____

13. How is the speed of a dc motor controlled *below* the base speed? _____
    _____
    _____

14.   How may an SCR replace a rheostat? _____

_____

15.   List the advantages of using thyristors in the motor drive control? _____

_____

_____

THE DC VOLTAGE DROP ACCELERATION CONTROLLER

OBJECTIVES

After studying this unit, the student will be able to

- state the purpose of a dc voltage drop acceleration controller.
- explain the principle of operation of lockout relays in a dc voltage drop acceleration controller.
- list, in sequence, the steps in the operation of an acceleration controller.
- connect an acceleration controller to a dc motor.

Large dc motors must be accelerated in controlled steps. A series of resistors or tapped resistors connected to lockout relays can be used to provide uniform motor acceleration. Since the starting current of a motor is high, the voltages across the series starting resistors are high and the voltage across the motor armature is low. As the motor accelerates and the counter emf increases, the armature current and the voltage drop across the series starting resistors decrease. Lockout relays connected across these resistors are calibrated to short out the series starting resistors as the motor speeds up.

CIRCUIT FOR THE VOLTAGE DROP ACCELERATION CONTROLLER

The Lockout Relay

For this circuit, motor acceleration is divided into three steps due to the actions of three lockout relays. A lockout relay has two coils, as shown in figure 9–1. For each relay shown in figure 9–2, coil A is connected across the line and serves as the operating coil of the relay. Coil LA of each relay is connected across one starting resistor. Since there are three starting resistors, three lockout relays are used to provide three steps of acceleration. The voltage drop across the resistors is high enough during the starting period to prevent the main control coil A of each relay from closing the A contacts which shunt or bypass the starting resistors. This action is called *lockout*.

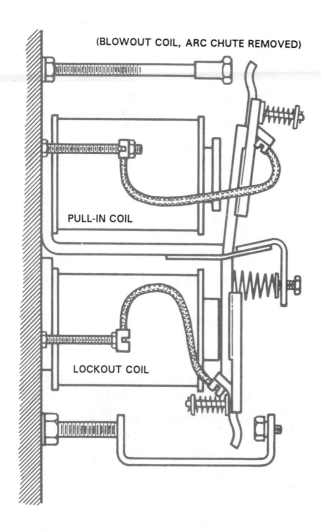

(BLOWOUT COIL, ARC CHUTE REMOVED)

PULL-IN COIL

LOCKOUT COIL

Fig. 9–1   A lockout relay

Starting the Motor

When the start button in figure 9–2 is pressed, the control relay M is energized. The main contacts 9-10 close and complete the armature circuit through the thermal element and the three resistors. The shunt field is connected across the line. Contacts 3–6 and 6–4 seal in the start button contact. Contacts 3–6 also energize coil 1A. The large value of starting current through the starting resistance produces a large voltage drop across each section of the starting resistance. As a result, there is a large voltage to the lockout coils 1 LA, 2 LA, and 3 LA and the accelerating contacts 1A, 2A, 3A are held open.

Contacts 4–6 are used to prevent coils 1A, 2A, and 3A from energizing when the start button is depressed. These coils receive power only when coil M is energized so they are not energized before coils 1LA, 2LA, and 3 LA.

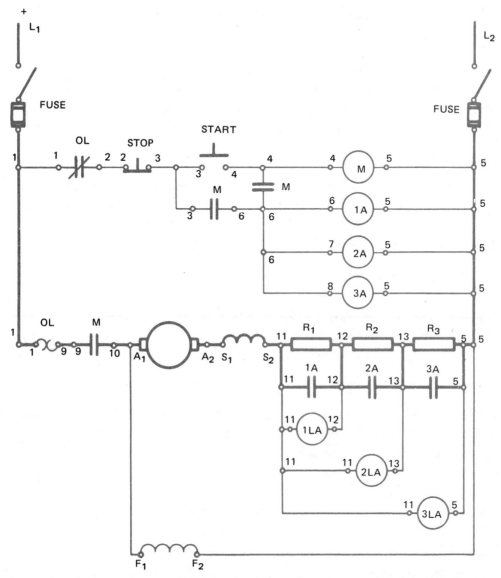

Fig. 9–2   Elementary diagram for the dc voltage drop acceleration controller

Acceleration

As the motor accelerates, the counter emf increases in the armature and the armature current decreases. With reduced current in the lockout coil of the relay, the pull of the lockout coil on the relay armature is less than the pull of the operating coil 1A. Therefore, relay 1A closes and contacts 1A bypass resistor R_1. Relay coil 2LA remains open due to the voltage drop across R_2, while relay coil 3LA remains open due to the voltage drop across R_2 and R_3. As R_1 is bypassed, the current again increases and the voltage drop across R_2 is high enough to keep the contacts of relay 2A open. However, as the motor continues to accelerate, the armature current decreases and lockout coil 2LA allows relay 2A to close contacts 2A and bypass resistor R_2.

Connecting the Motor across the Line

Resistor R_3 is bypassed in the same manner as R_1 and R_2. The motor is accelerated to normal speed in three steps. It is connected directly across the line when resistor R_3 is cut out.

Running Overload Protection

A sustained overload greater than the permissible percentage of the full-load current causes the thermal element to open overload contacts 1–2 which, in turn, causes coil M to become deenergized. Contacts 9-10 are then opened and the motor is shut down. Manual shut-down protection is obtained by pressing the stop button.

SUMMARY

The dc voltage drop acceleration controller can be used to provide definite steps of acceleration. The concept used is to electrically lock out the resistor shunting contacts. The large armature current operates a lockout portion of the relay and will not allow the shunting contacts to close until the armature current has decreased.

ACHIEVEMENT REVIEW

Select the correct answer or answers for each of the following statements, and place the corresponding letter or letters in the space provided.

1. High motor starting current will
   a. increase the field current.
   b. decrease the speed.
   c. increase the voltage drop across the starting resistors.
   d. decrease the voltage drop across the starting resistors.

2. The voltage drop acceleration method of speed control is particularly suited to
   a. braking of motors.
   b. starting small motors.
   c. starting large motors in one step.
   d. starting large motors in three steps.

3. Large dc motors are started with
   a. single-step acceleration.
   b. multiple-step acceleration.
   c. elementary counter emf controllers.
   d. auxiliary motors.

4. The lockout relay has _____
    a. one main coil.
    b. three coils.
    c. one closing and one lockout coil.
    d. two coils wound in series.

5. The main coil in a lockout relay is connected across the source
    and the lockout coil is energized by the _____
    a. counter emf.
    b. voltage drop across the starting resistor.
    c. total current in armature.
    d. current in the field circuit.

6. The lockout coil of the lockout relay weakens and causes
    the relay contacts to close when the _____
    a. motor speed drops.
    b. counter emf rises and reduces the motor current.
    c. counter emf decreases.
    d. starting resistance is cut out.

7. A decreasing starting current causes the lockout relay to _____
    a. cut in the starting resistance,
    b. cut out the starting resistance.
    c. increase the field resistance.
    d. decrease the field resistance.

8. Motor circuit protection during the starting period is
    provided by the _____
    a. thermal unit.
    b. branch fuses.
    c. distribution line fuses.
    d. motor circuit breaker.

9. A running load in excess of 150 percent of the full-load current
    causes the _____
    a. main line circuit to open before the control circuit.
    b. control circuit to be opened,
    c. speed to decrease.
    d. field rheostat to be bypassed.

10. The panel wiring is completed by referring to the _____
    a. relay symbols.   c. conduit plan.
    b. elementary diagram.   d. motor schematic diagram.

STUDENT ACTIVITY

Complete the conduit plan (figure 9–3) using the panel wiring diagram (figure 9–4) as a reference.

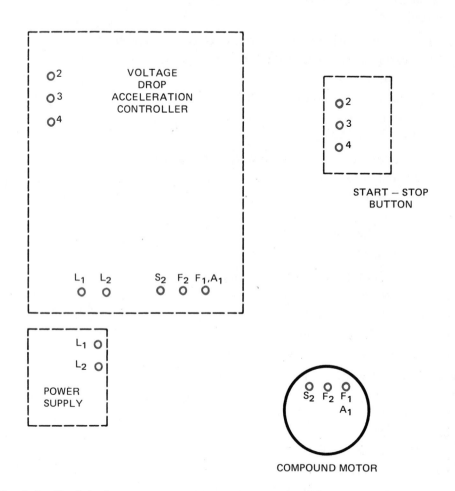

Fig. 9–3 Conduit plan, external wiring for a dc voltage drop acceleration controller

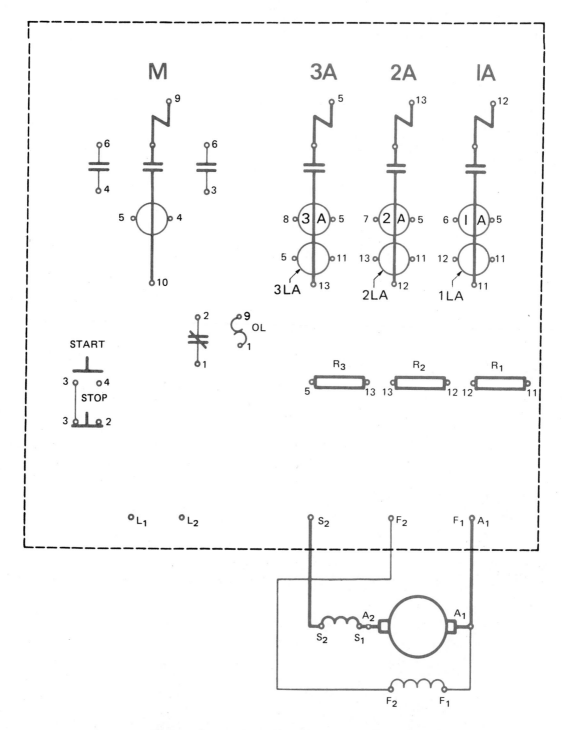

Fig. 9–4 Panel wiring diagram for the dc voltage acceleration controller with pushbuttons in the hinged cover

U•N•I•T
10

THE DC SERIES LOCKOUT RELAY ACCELERATION CONTROLLER

OBJECTIVES

After studying this unit, the student will be able to

- describe the operation of a dc series lockout relay.

- list the steps in the operating sequence of a series lockout relay acceleration controller.

A series lockout relay with its two coils is shown in figure 10–1. The lockout coil prevents the relay contacts from closing during the period of high motor starting current.

Note: In another type of relay, the magnetic gap is changed by shims located in back of the coil, rather than by an adjusting screw.

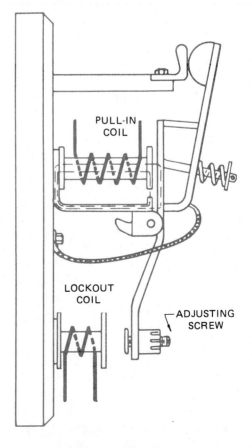

PULL-IN COIL

LOCKOUT COIL

ADJUSTING SCREW

Fig. 10–1   Series lockout relay

The pull-in coil closes the relay contacts when the motor has accelerated and the starting current is reduced.

OPERATION OF THE SERIES LOCKOUT ACCELERATION CONTROLLER

Figure 10–2 is the elementary diagram of a controller which uses series lockout relays to produce the necessary time delay or step control in motor acceleration.

Coils 1A, 2A, and 3A are the pull-in coils of the three relays, while coils 1HA, 2HA, and 3HA are the lockout coils of the relays.

When the start button is pressed, coil M is energized, and the main contactor, M, (1) seals the bypass auxiliary contact around the start button, and (2) allows current to flow in the following sequence: from L_1 through contact M, the motor armature, the series field, coil 1A, coil 1HA, resistors R_1, R_2, and R_3, and coil 3HA to L_2.

The large starting current through coil 1HA produces a large magnetic effect in this coil. Since this effect is larger than that of coil 1A, the relay is held open due to the fact that the magnetic path of the pull-in coil has a small amount of iron. As a result, this coil becomes saturated at high values of current. The magnetic circuit of the lockout coil has

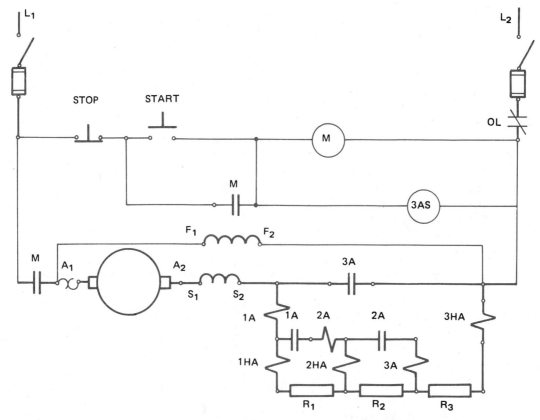

Fig. 10–2   Elementary diagram for the series lockout acceleration controller

a larger amount of iron and, therefore, does not tend to become saturated at a high current value.

As the motor accelerates and the starting current decreases, the pull of the lockout coil becomes less than the pull of the pull-in coil and contactor 1A closes. Resistor R_1 and lockout coil 1HA are bypassed and thus increased current is allowed through lockout rela coils 2A and 2HA. A cycle of operation similar to that which occurred for relay 1A now takes place for relay 2A. The current path is now from Li through M, the armature and series field, coil 1A, contact 1A, coils 2A and 2HA, resistors R_2 and R_3, and coil 3HA to L_2. The large current through lockout coil 2HA produces more magnetic pull than is present at coil 2A with the result that contact 2A is held open.

Finally, as the motor accelerates, all resistors and relays are shunted out. This should immediately cause all relay contacts to drop out, except for the fact that coil 3AS is an auxiliary shunt coil which acts on contactor 3A. Coil 3AS is strong enough to hold contact 3A closed after it has been pulled into contact, but is not strong enough to cause 3A to close its contacts without the aid of coil 3A.

When there is a heavier load on the motor, acceleration takes place over a longer period of time.

SUMMARY

The operation of the series lockout relay differs from the voltage drop relay. The lockout relay coil is in series with the pull-in coil. The magnetic fields are designed to pull the lockout coils first. As the current to the armature decreases, the lockout coil field is weakened and the shunting coils are allowed to pull the contacts shut, bypassing the armature resistors.

ACHIEVEMENT REVIEW

1. How is control relay M in the series lockout acceleration controller held in when the start button is released?

   _____

   _____

2. How does the lockout relay short out a starting resistor when the motor is accelerated?

   _____

   _____

3. What will happen if a break occurs in the coil of 3A?

   _____

   _____

4. A motor accelerates through two steps only. The speed is below normal and the
    motor voltage is low at the armature terminals. What is the probable cause of the
    problem?

    _____

    _____

    _____

5. What is the purpose of the 3AS coil of the lockout relay?

    _____

    _____

    _____

11

DYNAMIC BRAKING WITH A DC MOTOR REVERSAL CONTROL

OBJECTIVES

After studying this unit, the student will be able to

- list the steps in the operation of a dc motor control with interlocked forward and reverse pushbuttons.
- explain the principle of dynamic braking.
- describe the operation of a counter emf motor controller with dynamic braking.

Industrial motor installations often require that motors be stopped quickly and that the direction of rotation be reversed immediately after stopping. To achieve this operation, electrically and mechanically interlocked pushbutton stations connected to relays are used to disconnect the armature from the supply source. The armature is then connected to a low value of resistance. Because the inertia of the armature and connected load causes the armature to continue to revolve, it acts as a loaded generator. As a result, the armature is slowed in speed. This action is *called dynamic braking.*

Reversal of motors and dynamic braking are operations used in special equipment such as cranes, hoists, railway cars, and elevators.

MOTOR REVERSAL CONTROL

A motor is reversed by reversing the armature connections. The type of compounding is not affected by this method of obtaining reversal.

The pushbutton control station illustrated in figure 11–1 is the type used for motor reversal. The forward and reverse buttons are mechanically interlocked so that it is not possible to operate these buttons at the same time.

Description of Operation

Forward Starting. When the forward button is pressed, the normally open forward contacts close and the normally closed forward contacts open. The control circuit is shown in figure 11–2. The forward contactor coil is energized from L_1 through the overload contacts, stop button, forward pushbutton contacts 1–2 (when closed), and reverse

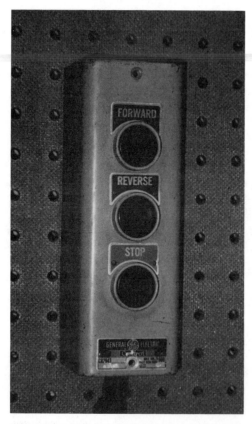

Fig. 11–1   A forward, reverse, stop pushbutton station

button contacts 3-4 through the forward contactor coil to L_2. The forward contacts F seal in the forward pushbutton. In the power circuit (figure 11–2) the F contacts of the forward energized contactor close, and thus complete the armature circuit through the starting resistance. The normal counter emf starter sequence of operations then continues to completion.

Reverse Operation. If the reverse pushbutton is pressed, contacts 3–4 of the reverse button open, and thus deenergize the forward contactor coil F. In addition, the F contacts are opened as well as the sealing contacts F. Pressing the reverse button also completes the circuit of the reverse contactor coil R which closes the R contacts. The motor armature circuit is now complete from L_1 to A_2 and A_1 to L_2 (figure 11–2). The armature connections are reversed and the armature rotates in the opposite direction. It is impossible for the reverse contacts to close until the forward contacts are open, due to the electrical and mechanical interlocking system used in this type of control circuit. The mechanical interlocks are shown by the broken lines between the R and F coils in figure 11–2.

Dynamic Braking

The purpose of dynamic braking is to bring a motor to a quicker stop. To do this there must be a method to quickly use the mechanical energy stored in the momentum of

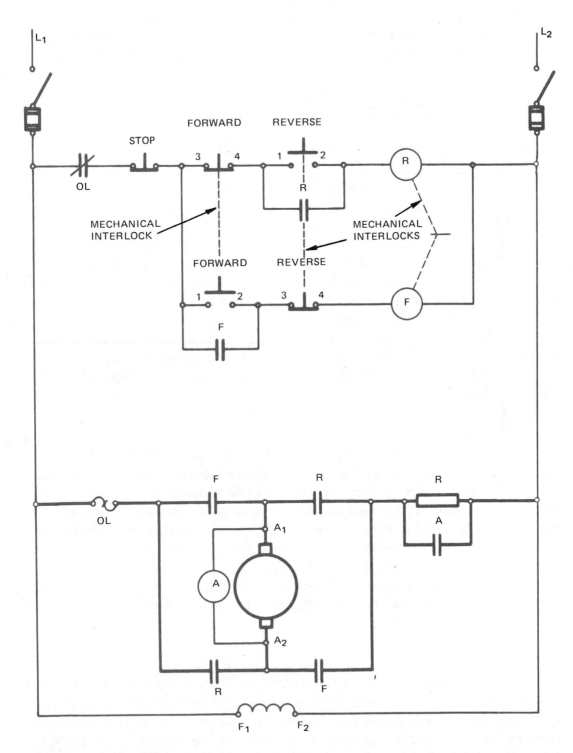

Fig. 11–2 Electrically and mechanically interlocked control and powered circuit for reversing motor

the armature after the main switch is opened. One method is to change the function of the motor to that of a generator. (A generator converts mechanical energy into electrical energy.) At the instant the motor is disconnected from the line, a resistor is connected across the motor armature. The resistor loads the motor as a generator, dissipates the mechanical energy, and slows the motor quickly.

DYNAMIC BRAKING USED IN A COUNTER EMF CONTROLLER

As an example, the principle of dynamic braking is shown by following the steps in the operation of an elementary counter emf controller (figure 11–3).

This analysis emphasizes the dynamic braking operation rather than the details of the circuit which were presented previously. The dynamic braking coil (DBM) is designed so that its only function is to insure a positive closing of the normally closed dynamic braking contacts 9–10. If the main coil M is energized, the dynamic braking contacts 9–10 open and contacts 8–9 of M close, although the dynamic braking coil is also energized. The dynamic braking coil is a little weaker than the M coil.

When the start button is pressed, the control coil M is energized, contacts 8–9 of M close, and the motor starts and accelerates up to normal speed by the counter emf method. At the instant the M control relay is energized, the main, normally closed, dynamic braking contacts 9–10 open. As a result, the dynamic brake resistor connection across the armature is broken.

Field Discharge Resistor

When using the counter emf controller *and* dynamic braking, a field discharge resistor must be added. The shunt field is disconnected from the supply voltage and its magnetic field begins to collapse. The quick collapse of the magnetic field produces a very large "inductive kick" voltage–thousands of volts: If it does not have a discharge path, the high voltage actually begins to break down the field winding insulation. Normally the discharge path is through the armature, which allows a slower collapse and keeps the voltage small. A field discharge resistor is a thyristor-type device that conducts when the voltage across it is high enough, but has a high resistance to normal line voltage. The collapsing magnetic field's voltage can discharge through the FDR (field discharge resistor) without damaging the field windage.

Stopping

When the stop button is pressed, relay control coil M is deenergized, and M contacts 8–9 open the armature circuit and close the dynamic braking contacts 9–10. These contacts connect the dynamic brake resistor directly across the armature. Since the shunt field is still connected across the line and receiving full excitation, the high counter emf generated in the armature causes a high load current through the dynamic brake resistor. The heavy load

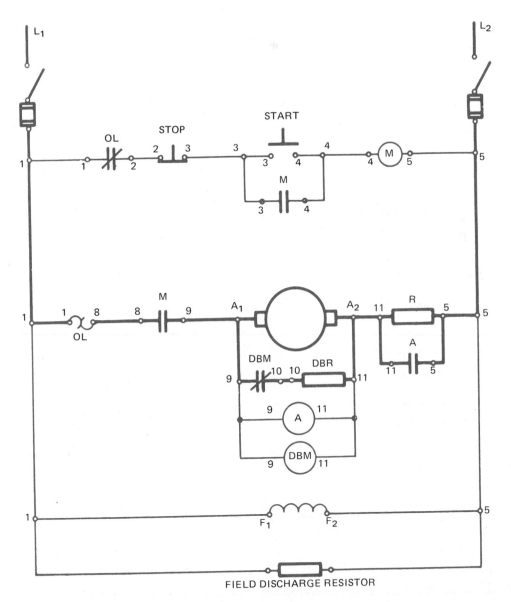

Fig. 11–3   Elementary diagram for a dc counter emf controller

current dissipates the stored mechanical energy in the armature with the result that the motor slows to a stop. File braking action decreases as the armature speed decreases.

SUMMARY

Figure 11–2 shows the reversing circuit with electrical pushbutton interlocks and mechanical interlocks between the forward and reverse contactor. The motor is reversed by reversing current flow through the armature, but keeping the shunt field current in the same direction. Dynamic braking is used when slowing the motor. As the armature is disconnected from the power source, a dynamic braking resistor is connected across the armature. The armature is still spinning and the shunt field is still energized, so the armature acts as a generator. The resistor provides a current path for the generated current and slows the armature as it works as a generator. A field discharge resistor is also used to prevent the sudden collapse of the shunt field flux as it is disconnected from the line power.

ACHIEVEMENT REVIEW

1. How can dc motors be reversed without changing the type of compounding?

   _____

2. What interlocking is necessary on the forward and reverse pushbuttons to avoid short circuits? _____

   _____

3. What will happen if the forward and reverse relays are energized at the same time?

   _____

   _____

4. How many contacts are required on the forward relay? _____

5. What are two applications requiring motor reversal? _____

   _____

6. How can a motor be shut down quickly without using a mechanical brake?

   _____

7. When is dynamic braking applied? _____

   _____

8. How does dynamic braking slow a motor? _____

   _____

   _____

9. If an electromagnetically operated brake is connected in series with the armature, how is it operated? _____

   _____

   _____

10. What are two installations where dynamic braking is used? _____

    _____

    _____

12

SUMMARY REVIEW
OF UNITS 5-11

OBJECTIVE

- To give the student an opportunity to evaluate the knowledge and understanding acquired in the study of the previous seven units.

A. Complete the following statements.

1. A dc motor starter is designed to limit the _____ .

2. A three-terminal starting rheostat provides _____ protection for a shunt motor.

3. Since the holding coil of a four-terminal rheostat is connected across the source, this type of rheostat provides _____ protection to a motor.

4. A four-terminal rheostat may be used with a motor for which a wide range of _____ control is required.

5. Series motor starters provide either _____ protection or _____ protection.

6. A drum controller is used when the operator has _____ control of the motor.

B. Select the correct answer for each of the following statements and place the corresponding letter in the space provided.

7. The three-terminal starting box can be used with all but the _____
   a. compound motor.   c. series motor.
   b. shunt motor.

8. Interpretation of automatic circuits requires the recognition of _____
   a. color.   c. parallel circuits.
   b. rating.   d. electrical circuit symbols.

9. A piece of apparatus which contains a minimum of one set of contacts operated by a coil is called a _____
   a. motor.   c. relay.
   b. magnet.   d. dynamo.

10. Normally closed contacts are   _____
    a. open at all times.
    b. open when the relay coil is deenergized.
    c. open when the relay coil is energized.
    d. closed when the relay coil is energized.

11. An elementary automatic motor controller circuit diagram
    shows the   _____
    a. actual wiring layout.
    b. schematic motor diagram.
    c. actual sequence of operations of the entire circuit.
    d. sequence of operations of the starting circuit only.

12. One type of automatic controller operates on the basis that
    the counter emf generated in a motor   _____
    a. increases the starting current.
    b. energizes a starting relay.
    c. deenergizes a starting relay.
    d. reduces the field current.

13. Another type of controller accelerates the motor in steps by   _____
    a. using a series of thermal elements.
    b. allowing starting resistor voltage drops to energize relays.
    c. operation of relays controlled by speed.
    d. cam controlled relays.

14. A controller is called a series lockout relay starter because the   _____
    a. starting resistors are bypassed and locked out.
    b. relays are all connected in series.
    c. relay coils are connected in series with the armature and
       cause each relay to lock out (open).
    d. relay coils lock themselves out of the circuit.

15. A remote-controlled motor reversal requires   _____
    a. one relay with two contacts.
    b. two single contactor relays.
    c. two double contactor relays.
    d. two interlocked relays.

16. The principle of dynamic braking involves   _____
    a. the use of a dynamo as a brake.
    b. the use of an electromagnetic brake connected across
       the motor armature.
    c. connecting a low resistance across the motor armature.
    d. connecting a brake relay across the armature.

C. Complete each of the statements at the left by selecting the letter of the appropriate phrase or phrases from the list at the right. Write the letter(s) in the space provided.

17. Undervoltage protection is provided in

18. Dynamic braking is used for

19. Multiple-step acceleration is used for starting

20. Automatic starters provide more uniform acceleration than

21. A sealing circuit is used in

a. quick stopping.
b. large motors.
c. a three-terminal starting rheostat.
d. a four-terminal starting rheostat.
e. small motors.
f. series motors.
g. manual starters.
h. control circuits.

22. Match each of the following symbols with its description at the right. Place the letter of the symbol in the space provided.

a. $\dashv\vdash$

b.

c. $-\!\!\bigcirc\!\!\!{\scriptstyle M}\!\!-$

d.

e.

f.

g. $-\!\Box\!-$

1. _____ Normally closed pushbutton

2. _____ Thermal motor overload relay

3. _____ Shunt relay coil

4. _____ Normally open contact

5. _____ Fuse

6. _____ Normally closed contact

7. _____ Normally open pushbutton

13

ELECTROMECHANICAL AND SOLID-STATE RELAYS AND TIMERS

OBJECTIVES

After studying this unit, the student will be able to

- explain how relays operate.
- list the principal uses of relays.
- describe different relay control and load conditions.
- tell how SCRs operate.
- identify relay component symbols.
- connect different relays in a circuit.
- identify and use various timers.
- use proper timer symbols in schematic diagrams.

RELAYS

Relays are devices used to relay or multiply electrical contact closures. The relay concept is used where a small voltage at low current is used to operate a set of electrical contacts to an open or closed position. This contact operation in turn controls a larger electric load and so on, as it relays the electrical operations.

Another common use of a relay is to multiply a single signal to open or close multiple contacts for control of multiple electrical loads.

ELECTROMECHANICAL RELAYS

Electromechanical relays, contactors, and motor starters basically operate by the same principles. These electrically-operated switches respond to the electromagnetic attraction of an energized coil of wire in an iron core. The devices differ in the amount of current that each must switch. The relay—which can be compared to an amplifier—is used to switch small amounts of currents (usually 0–15 amperes) in many control circuits (figure

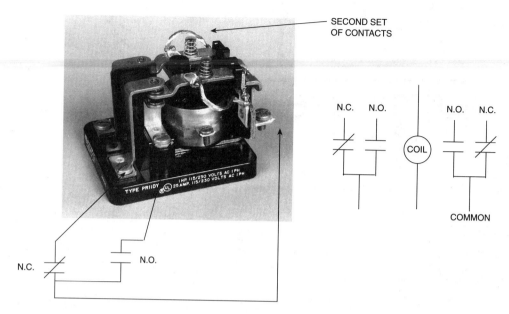

Fig. 13–1   Control relay and associated coil contact diagram (From Keljik, *Electric Motors and Motor Controls*, copyright 1995 by Delmar Publishers)

13–1). Uses of relays include: switching (on and off) larger coils of motor starters, contactors, solenoids, heating elements, and small motors. Other uses are alarm systems and pilot light control. Relays have many industrial and commercial applications, both ac and dc.

A small current flow and/or low voltage applied to a relay coil can result in a much larger current or voltage being switched. One input signal (voltage) may control several output (switched) circuits (figure 13–2).

The coil voltages of the relay are separate or different from those at the switched contacts; this is called *separate control*. However, the coil voltage may be the same system voltage as the switched voltage.

Relays are available in many shapes and sizes. Some are sealed in dust proof, transparent plastic enclosures (figure 13–3A). The general construction of a typical relay is shown in figure 13–3B.

Note in figure 13–2 that relay contacts may be normally-closed or normally-open. The action of the contacts is to switch something *on* or *off* depending on the configuration.

SOLID STATE RELAY (SSR)

A solid state relay can be used to control most of the circuits that the electromechanical relay controls. By comparison, the solid-state relay has no coil or contacts. The semiconductor industry has developed solid-state components with unusual applications to the industrial control processes. These components are very compact and versatile. They are very reliable if used in the proper application.

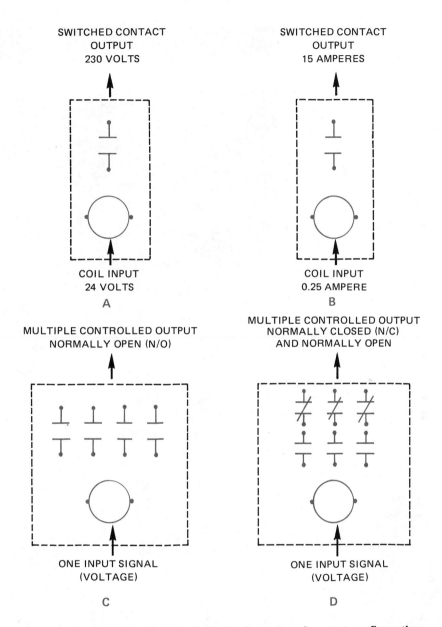

Fig. 13–2   Several electrical-mechanical relay uses and contact configurations

The silicon-controlled rectifier (SCR) is probably the most popular solid-state device for controlling large and small electrical power loads. Basically, the SCR is a rectifier which either conducts or does not conduct an electric current. When it is not conducting, the SCR offers almost a complete blockage to the current. It passes only a few milliamperes to the load. For this reason, some manufacturers place contacts from electrically operated contactors in the total circuit to disconnect the load completely.

Fig. 13–3A "Ice cube relays" and plug in bases used for easy replacement (From Keljik, *Electric Motors and Motor Controls*, copyright 1995 by Delmar Publishers)

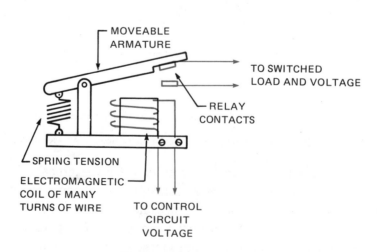

A

B

Fig. 13–3B Construction of a typical electromagnetic control relay

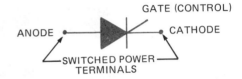

Fig. 13–4   Schematic symbol for an SCR which is the heart of the solid-state relay

The SCR will not conduct when the voltage across it is in the reverse direction. It will conduct only in the forward direction when the proper signal (voltage) is applied to the gate terminal (figure 13–4). Once it is conducting, the SCR cannot be turned off immediately. It is necessary only to provide a small signal to start the SCR conducting a current. It will continue to conduct even without a signal from that point on as long as the current is in the forward direction. The only way to stop the SCR from conducting is to reduce the current flow below the holding current level, or disconnect it from the system. On alternating current, of course, this happens every half-cycle, so this characteristic is no problem (figure 13–5). For direct-current applications, the voltage is reduced to zero by interrupting the circuit, generally with a contact on an electromagnetic relay.

Figure 13–6 shows a typical solid-state relay. Note the input and switched output terminal connections. Figure 13–7 shows these connections completed. Terminal wiring is very simple and consists of two input control wires and two output load wires. The connecting terminals are clearly identified on solid-state relays (as they are on electromechanical relays). The relay in figure 13–7 has a light-emitting diode (LED) connected to the input or control voltage. When the input voltage turns the LED on, a photodetector connected to the gate of the triac turns the triac on and connects the load to the line. This optical coupling is commonly used with solid-state relays. These relays are referred to as being *optoisolated*. This means that the load side of the relay is optically isolated from the control side of the relay. The control medium is a light beam. No voltage spikes or electrical noise produced on the load side of the relay are therefore transmitted to the control side.

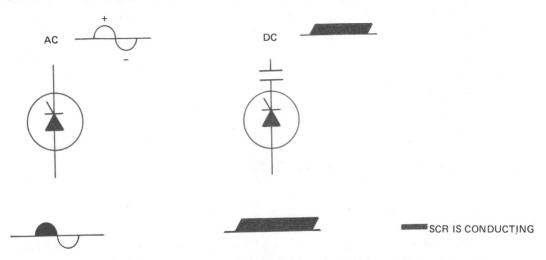

Fig. 13–5   An SCR will contact current in the forward direction until the voltage is reduced to zero

Fig. 13–6   Solid state relay has no moving parts (From Keljik, *Electric Motors and Motor Controls*, copyright 1995 by Delmar Publishers)

Solid state relays have a wide range of input or control voltage designs. The solid state relays may operate at TTL (transistor logic levels) + 5 volts dc. They may operate at levels between 3–30V dc or other relays at 90–120V ac. Considerations as to what style of relay include the way that the SSR turn on. A *zero switch* relay turns on as the load voltage crosses zero. This is often used for resistive type loads.

Instant-on relays turn on anytime during the ac waveform to the load. A third style is the *universal switch*. This turn-on occurs after the load voltage has reached peak and is moving toward zero. The universal switch is used for combination loads of inductive and resistive loads.

TIMERS

Timers come in many different styles and with different operating characteristics. The mechanical timers use either clock motors to operate a mechanical trip mechanism or solenoids to create timing operations. Pneumatic timers use air, a diaphragm, and an operating solenoid to cause a time delay. Figure 13–8 shows a mechanical time-clock-type timer. This type of timer is used for rough time-of-day timing to turn lights or equipment on or off at an approximate time of day. Tabs on the face of the clock are moved to create different on and off periods.

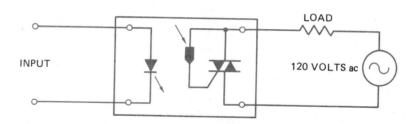

Fig. 13–7   Solid state relay used to control an ac load (From Herman, *Electronics for Industrial Electricians*, 2nd Edition, copyright 1989 by Delmar Publishers)

Fig. 13–8   Synchronous clock timer (*Courtesy of Wesco*. From Keljik, *Electric Motors and Motor Controls*, copyright 1995 by Delmar Publishers)

Pneumatic timers such as those in figure 13–9 are typically used for on or off delay operations. The symbols shown in figure 13–10 depict the type of timer operation and the operation of the contacts. In ON-delay timers (TDE, Time-Delay-on-Energization) the contacts stay in their original position until a solenoid plunger has moved through its entire travel distance. The travel time is controlled by adjusting a needle valve to allow the air to escape in front of the solenoid diaphragm. The timer solenoid has power applied to pull the plunger into its timed-out position. The contacts can start as either normally-open or normally-closed and will change to the opposite position at the end of the time, if power is still applied to the solenoid coil.

The OFF-delay timer (TDD, Time-Delay-on-De-energization) acts in the opposite mode. In other words, the timing change to the contacts takes place when the power is removed from the timer coil and the solenoid plunge is allowed to go back to it de-ener-

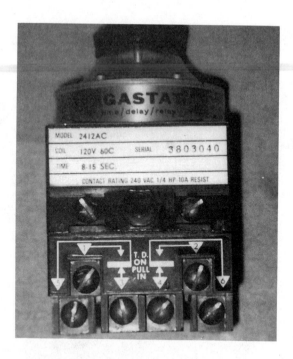

Fig. 13–9   Time delay relay symbols used with timers (From Keljik, *Electric Motors and Motor Controls*, copyright 1995 by Delmar Publishers)

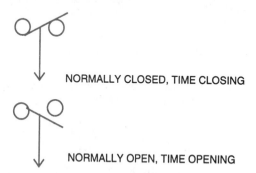

NORMALLY CLOSED, TIME CLOSING

NORMALLY OPEN, TIME OPENING

Fig. 13–10   Time delay on de-energization contact symbols

gized position. The time that it takes the solenoid to return to it original position is controlled by the needle valve allowing air to return to the vacuum side of the air diaphragm. Figure 13–10 shows the TDD contacts in their normal, timed-out state. When power is applied to the timing coil in this type timer, the contacts change to the opposite state and will time out to return to the original designation.

Fig. 13–11 Electronic timers with side range of functions and time delays (From Keljik, *Electric Motors and Motor Controls*, copyright 1995 by Delmar Publishers)

Solid state timers are now commonly used to provide highly accurate and an extremely wide range of timing operations. Many of the timers, such as the one in figure 13–11, can be used as ON-delays or OFF-delays and have several different trigger modes. They can also provide a wide range of timing, from milliseconds to hundreds of hours. The timing is accomplished with electronics, which makes them very accurate and precise.

SUMMARY

Relays are used to control various loads from other electrical circuits. They may be used to operate large values of dc or ac power or simply multiply the electrical circuit function. Contacts can be normally-open (NO) or normally-closed (NC) or convertible from one form to another.

Things to consider when ordering:

Control voltage and value;

Contact ratings in current and voltage, number of contacts, NO, NC, or SPDT, etc.

Are open contacts, enclosed contacts, or solid-state no-contact movement important?

Does the relay need very fast speed as in SSR or is physical isolation more important as in electromechanical relays?

ACHIEVEMENT REVIEW

1. A mouse trap may be compared with the gate trigger action of an SCR. Indicate whether this statement is true or false and explain your reasoning. _____

   _____

   _____

   _____

   _____

2.   What is the major difference between an electromechanical relay and a solid-state
     relay? _____

     _____

     _____

     _____

3.   Describe different relay control and load conditions. _____

     _____

     _____

     _____

4.   How are relays used in industrial controls?_____

     _____

     _____

     _____

5.   Basically, what is an SCR? _____

     _____

     _____

     _____

     _____

     _____

     _____

14

THREE-PHASE, SQUIRREL-CAGE INDUCTION MOTOR

OBJECTIVES

After studying this unit, the student will be able to

- describe the construction of a three-phase, squirrel-cage motor, listing the main components of this type of motor.
- identify the following items and explain their importance to the operation of a three-phase, squirrel-cage induction motor: rotating stator field, synchronous speed, rotor induced voltages, speed regulation, percent slip, torque, starting current, no-load power factor, full-load power factor, reverse rotation, and speed control.
- calculate motor speed and percent slip.
- reverse a squirrel cage motor.
- describe why a motor draws more current when loaded.
- draw diagrams showing the dual voltage connections for 230/460-volt motor operation.
- explain motor nameplate information.

OPERATING CHARACTERISTICS

The three-phase, squirrel-cage induction motor is relatively small in physical size for a given horsepower rating when compared with other types of motors. The squirrel-cage induction motor has very good speed regulation under varying load conditions. Because of its rugged construction and reliable operation, the three-phase, squirrel-cage induction motor is widely used for many industrial applications (Figure 14–1).

CONSTRUCTION DETAILS

The three-phase, squirrel-cage induction motor normally consists of a stator, a rotor, and two end shields housing the bearings that support the rotor shaft.

A minimum of maintenance is required with this type of motor because

- the rotor windings are shorted to form a squirrel cage.

Fig. 14–1   Three phase motors used for pumping application

- there are no commutator or slip rings to service (compared to the dc motor).
- there are no brushes to replace.

The motor frame is usually made of cast steel. The stator core is pressed directly into the frame. The two end shields housing the bearings are bolted to the cast steel frame. The bearings which support the rotor shaft are either sleeve bearings or ball bearings. Figure 14–2 is a cutaway view of an assembled motor. Figure 14–3 illustrates the main parts of a three-phase, squirrel-cage induction motor.

Stator

A typical stator contains a three-phase winding mounted in the slots of a laminated steel core (Figure 14–4). The winding itself consists of formed coils of wire connected so

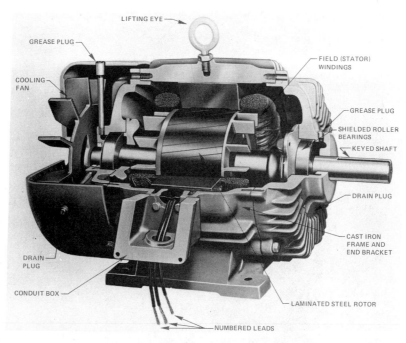

Fig. 14–2 Cutaway view of construction and features of a typical three-phase explosionproof motor (*Photo courtesy of Marathon Electric Manufacturing Corp*)

Fig. 14–3 Main components of a squirrel-cage induction motor (*Courtesy General Electric Company*)

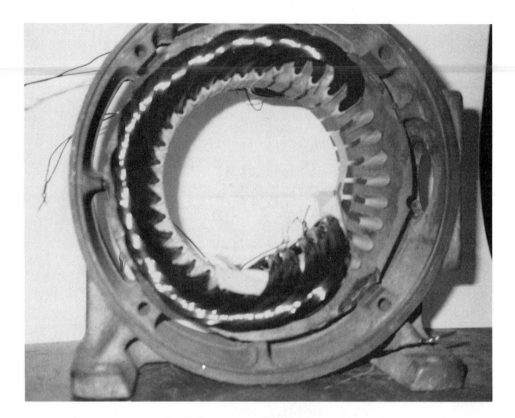

Fig. 14–4   Partially wound stator of three phase motor

that there are three single-phase windings spaced 120 electrical degrees apart. The three separate single-phase windings are then connected, usually internally, in either wye or delta. Three or nine leads from the three-phase stator windings are brought out to a terminal box mounted on the frame of the motor for single- or dual-voltage connections.

Rotor

The revolving part of the motor consists of steel punchings or laminations arranged in a cylindrical core (figure 14–5 to 14–7). Copper or aluminum bars are mounted near the surface of the rotor. The bars are brazed or welded to two copper end rings. In some small squirrel-cage induction motors, the bars and end rings are cast in one piece from aluminum.

Figure 14–5 shows such a rotor. Note that fins are cast into the rotor to circulate air and cool the motor while it is running. Note also that the rotor bars between the rings are skewed at an angle to the faces of the rings. Because of this design, the running motor will be quieter and smoother in operation. A keyway is visible on the left end of the shaft. A pulley or load shaft coupling can be secured using this keyway.

Fig. 14–5 Squirrel cage rotor for an induction motor (*Courtesy General Electric Company*)

Fig. 14–6 Cutaway view of a cage rotor (*Courtesy General Electric Company*)

Fig. 14–7 Squirrel-cage form for an induction motor (*Courtesy General Electric Company*)

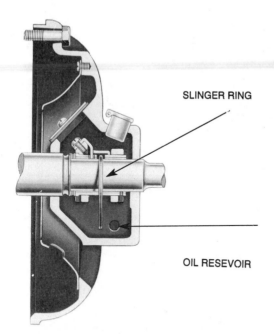

SLINGER RING

OIL RESEVOIR

Fig. 14–8   Sleeve-bearing end shield for an open polyphase motor (*Courtesy General Electric Company*)

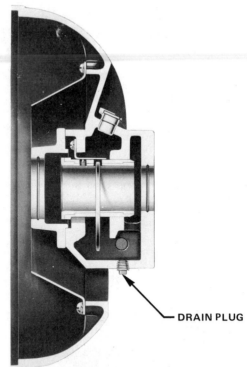

DRAIN PLUG

Fig. 14–9   Sleeve-bearing end shield for a polyphase induction motor (*Courtesy General Electric Company*)

Shaft Bearings

Typical sleeve bearings are shown in figures 14–8 and 14–9. The inside walls of the sleeve bearings are made of a babbitt metal which provides a smooth, polished, and long wearing surface for the rotor shaft. A large oversized oil slinger ring fits loosely around the rotor shaft and extends down into the oil reservoir. This ring picks up and slings oil on the rotating shaft and bearing surfaces. Two oil rings are shown in figure 14–10. This lubricating oil film minimizes friction losses. An oil inspection cup on the side of each end shield enables maintenance personnel to check the level of the oil in the sleeve bearing.

Figures 14–11 to 14–14 illustrate ball bearing units. In some motors, ball bearings are used instead of sleeve bearings. Grease rather than oil is used to lubricate ball bearings. This type of bearing usually is two-thirds full of grease at the time the motor is assembled. Special fittings are provided on the end bells so that a grease gun can be used to apply additional lubricant to the ball bearing units at periodic intervals.

When lubricating roller bearings, remove the bottom plug so that the old grease is forced out. The manufacturer's specifications for the motor should be consulted for the lubricant grade recommended, the lubrication procedure, and the bearing loads.

Fig. 14–10 Partially assembled sleeve bearing for a totally enclosed, 1,250-hp motor (*Photo courtesy of Siemans-Allis*)

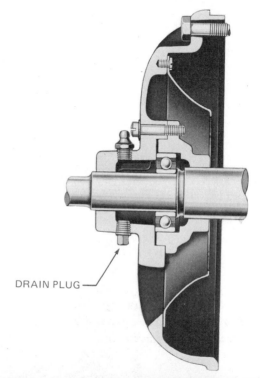

DRAIN PLUG

Fig. 14–11 Ball bearing end shield for an open polyphase motor (*Photo courtesy of General Electric Company*)

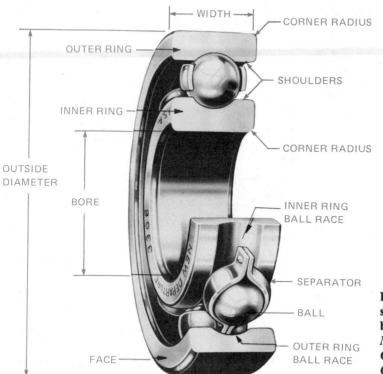

WIDTH

CORNER RADIUS

OUTER RING

SHOULDERS

INNER RING

CORNER RADIUS

OUTSIDE
DIAMETER

BORE

INNER RING
BALL RACE

SEPARATOR

BALL

OUTER RING
BALL RACE

FACE

**Fig. 14–12   Cutaway
section of a single-row ball
bearing (*Photo courtesy of
New Departure Division,
General Motors
Corporation*)**

**Fig. 14–13   Single, sealed-type ball bearing
(*Photo courtesy of New Departure Division,
General Motors Corporation*)**

Fig. 14–14   Double-row ball bearing *(Photo courtesy of New Departure Division, General Motors Corporation)*

PRINCIPLE OF OPERATION OF A SQUIRREL-CAGE MOTOR

As stated in a previous paragraph on the stator construction, the slots of the stator core contain three separate single-phase windings. When three currents 120 electrical degrees apart pass through these windings, a rotating magnetic field results. This field travels around the inside of the stator core. The speed of the rotating magnetic field depends on the number of stator poles and the frequency of the power source. This speed is called the *synchronous speed* and is determined by the formula:

$$\text{Synchronous speed RPM} = \frac{120 \times \text{frequency in hertz}}{\text{Number of poles}}$$

$$S = \frac{120 \times f}{p}$$

S = Synchronous speed

f = Hertz (frequency)

p = Number of poles per phase

Example 1. If a three-phase, squirrel-cage induction motor has six poles on the stator winding and is connected to a three-phase, 60-hertz source, then the synchronous speed of the revolving field is 1,200 RPM-Revolutions Per Minute .

$$S = \frac{120 \times f}{p} = \frac{120 \times 60}{6} = 1{,}200 \text{ RPM}$$

As this magnetic field rotates at synchronous speed, it cuts the copper bars of the rotor and induces voltages in the bars of the squirrel-cage winding. These induced voltages set up currents in the rotor bars which in turn create a field in the rotor core. This rotor field reacts with the stator field to cause a twisting effect or torque which turns the rotor. The rotor always turns at a speed slightly less than the synchronous speed of the stator field. This means that the stator field will always cut the rotor bars. If the rotor turns at the same speed as the stator field, the stato field will not cut the rotor bars and there will be no induced voltage or torque.

Speed Regulation and Percent Slip

The squirrel-cage induction motor has very good speed regulation characteristics (the ratio of difference in speed from no load to full load). Speed performance is measured in terms of percent slip. The synchronous speed of the rotating field of the stator is used as a reference point. Recall that the synchronous speed depends on the number of stator poles and the operating frequency. Since these two quantities remain constant, the synchronous speed also remains constant. If the speed of the rotor at full load is deducted from the synchronous speed of the stator field, the difference is the number of revolutions per minute that the rotor slips behind the rotating field of the stator.

$$\text{Percent Slip} = \frac{\text{synchronous speed} - \text{rotor speed}}{\text{synchronous speed}} \times 100$$

Example 2. If the three-phase, squirrel-cage induction motor used in Example 1 has a synchronous speed of 1,200 r/min and a full-load speed of 1,140 r/min, find the percent of slip.

Synchronous speed (Example 1)   = 1,200 RPM
Full-load rotor speed   = 1,140 RPM

$$\text{Percent slip} = \frac{\text{synchronous speed} - \text{rotor speed}}{\text{synchronous speed}} \times 100$$

$$\text{Percent slip} = \frac{1{,}200 - 1{,}140}{1{,}200} \times 100$$

$$\text{Percent slip} = \frac{60}{1{,}200} \times 100 = .05 \times 100$$

$$\text{Percent slip} = 5 \text{ percent}$$

For a squirrel-cage induction motor, as the value of percent slip decreases toward 0%, the speed performance of the motor is improved. The average range of percent slip for squirrel-cage induction motors is 2 percent to 6 percent.

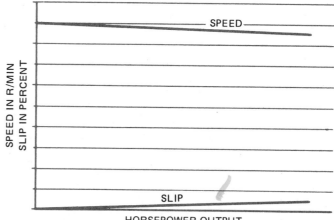

Fig. 14–15   Speed curve and percent slip curve

Figure 14–15 shows a speed curve and a percent slip for squirrel-cage induction motor operating between no load and full load. The rotor speed at no load, slips behind the synchronous speed of the rotating stator field just enough to create the torque required to overcome friction and windage losses at no load. As a mechanical load is applied to the motor shaft, the rotor tends to slow down. This means that the stator field (turning at a fixed speed) cuts the rotor bars a greater number of times in a given period. The induced voltages in the rotor bars increase, resulting in more current in the rotor bars and a stronger rotor field. There is a greater magnetic reaction between the stator and rotor fields which causes a stronger twisting effect or torque. This also increases stator current taken from the line. The motor is able to handle the increased mechanical load with very little decrease in the speed of the rotor.

Typical slip-torque curves for a squirrel-cage induction motor are shown in Figure 14–16. The torque output of the motor in pound-feet (lb.ft) increases as a straight line with an increase in the value of percent slip as the mechanical load is increased to the point of full load. Beyond full load, the torque curve bends and finally reaches a maximum point called the breakdown torque. If the motor is loaded beyond this point, there will be a corresponding decrease in torque until the point is reached where the motor stalls. However, all induction motors have some slip in order to function. Starting torque is not shown, but is approximately 300% of running torque.

Starting Current

When a three-phase, squirrel-cage induction motor is connected across the full line voltage, the starting surge of current momentarily reaches as high a value as 400% to 600% or more of the rated full-load current. At the moment the motor starts, the rotor is at a standstill. At this instant, therefore, the stator field cuts the rotor bars at a faster rate than when the rotor is turning. This means that there will be relatively high induced voltages in the rotor which will cause heavy rotor current. The resulting input current to the

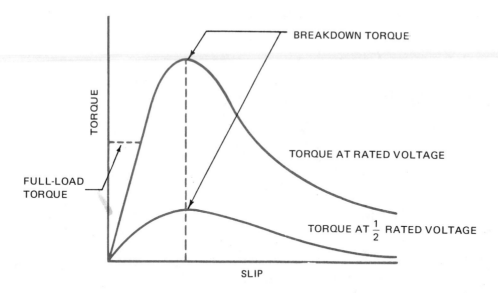

Fig. 14–16   Slip-torque curves for a running squirrel-cage motor

stator windings will be high at the instant of starting. Because of this high starting current, starting protection rated as high as 300 percent of the rated fullload current for nontime delay fuses is provided for squirrel-cage induction motor installations.

Most squirrel-cage induction motors are started at full voltage. If there are any questions concerning the starting of large sizes of motors at full voltage, the electric utility company should be consulted. In the event that the feeders and protective devices of the electric utility are unable to handle the large starting currents, reduced voltage starting circuits must be used with the motor.

Power Factor

The power factor of a squirrel-cage induction motor is poor at no-load and low-load conditions. At no load, the power factor can be as low as 15 percent lagging. However, as load is applied to the motor, the power factor increases. At the rated load, the power factor may be as high as 85 to 90 percent lagging.

The power factor at no load is low because the magnetizing component of input current is a large part of the total input current of the motor. When the load on the motor is increased, the in-phase current supplied to the motor increases, but the magnetizing component of current remains practically the same. This means that the resultant line current is more nearly in phase with the voltage and the power factor is improved when the motor is loaded, compared with an unloaded motor which has its magnetizing current as a major component of the input current.

Figure 14–17 shows the increase in power factor from a no-load condition to full load. In the no-load diagram, the in-phase current (I_w) is small when compared to the

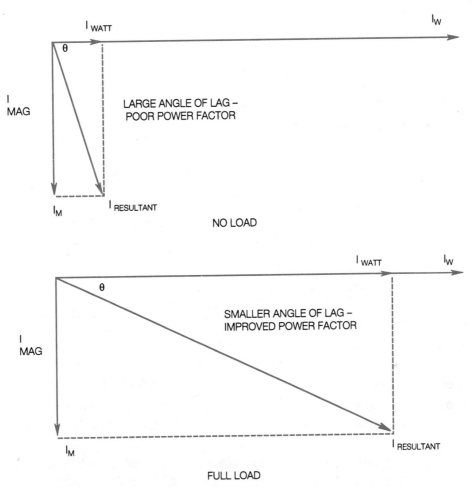

Fig. 14–17   Power factor at no load and full load

magnetizing current I_M; thus, the power factor is poor at no load. In the full-load diagram, the in-phase current has increased while the magnetizing current remains the same. As a result, the angle of lag of the line current decreases and the power factor increases.

Reversing Rotation

The direction of rotation of a three-phase induction motor can be reversed readily. The motor will rotate in the opposite direction if any two of the three line leads are reversed (Figure 14–18). The leads are reversed at the motor.

Speed Control

A squirrel-cage induction motor has almost no speed variations without external controls. Recall that the speed of the motor depends on the frequency of the three-phase source and the number of poles of the stator winding.

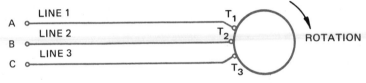

ROTATION BEFORE CONNECTIONS ARE CHANGED

**Fig. 14–18   Reversing
rotation of an induction motor**   ROTATION AFTER CONNECTIONS ARE CHANGED

The frequency of the supply line is usually 60 hertz, and is maintained at this value by the local power utility company. Since the number of poles in the motor is also a fixed value, the synchronous speed of the motor remains constant. As a result, it is not possible to obtain a range of speed without changing the applied frequency. It can be controlled by a variable-frequency ac electronic drive system,or by changing the number of poles using external controllers.

INDUCTION MOTORS WITH DUAL-VOLTAGE CONNECTIONS

Many three-phase, squirrel-cage induction motors are designed to operate at two different voltage ratings. For example, a typical dual-voltage rating for a three-phase motor is 230/460 volts.

Figure 14–19 shows a typical wye-connected stator winding which may be used for either 230 volts, three phase or 460 volts, three phase. Each of the three single-phase windings consist of two coil windings. There are nine leads brought out externally from this type of stator winding. These leads, identified as leads 1 to 9, end in the terminal box of the motor. To mark the terminals, start at the upper left-hand terminal T_1 and proceed in a clockwise direction in a spiral toward the center, marking each lead as indicated in the figure.

Figure 14–20 shows the connections required to operate a motor from a 460-volt, three-phase source. The two coils of each single-phase winding are connected in series, Figure 14–21 shows the connections to permit operation from a 230-volt, three-phase source.

Star Connected Motors

If the lead identification of a 9-lead (dual-voltage), 3-phase, Star connected motor have been destroyed, the electrician must reidentify them before connecting the motor to the line. The following method may be used. First, identify the internally connected star point by checking for continuity between three of the leads as in Figure 14–22 A.

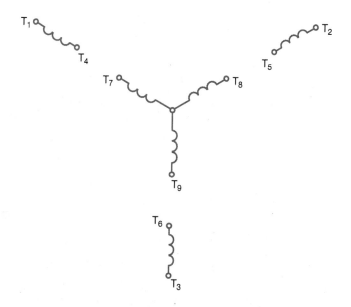

Fig. 14–19 Method of identifying terminal markings

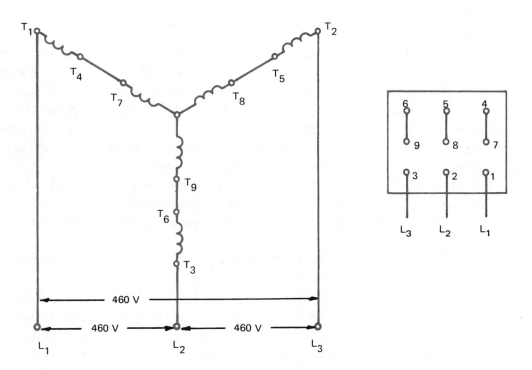

Fig. 14–20 460-volt wye connection. Coils are connected in series.

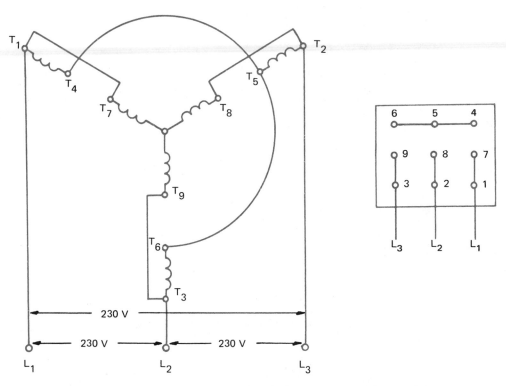

Fig. 14–21   230-volt wye connection. Coils are connected in parallel.

Then identify the three other sets of coils by continuity between two leads at a time (Figure 14–22 B). Assign T_7, T_8, and T_9 to any of the three leads of the permanent star connected coils (a). Apply the lower rated line voltage for the motor to T_7, T_8, and T_9 and operate to check the direction of rotation. Disconnect line voltage and connect one of the undetermined coils to T_7. Reconnect power, leaving the lines on T_7, T_8, and T_9. If the coil is correctly connected and is the proper coil, the voltage should be about 1.5 times the line voltage between the loose end and the other two lines. *Be careful of line voltage.*

If the correct coil is selected but reversed, the voltage between the loose end and the other two leads will be about 58% of the line voltage. If the wrong coil is selected, the voltage differences between the loose end and the other two line leads will be uneven (see Figure 14–22 C).

When the readings are even and approximately 1.5 times the line voltage, mark the lead connected to T_7 as T_4 and other end of the coil as T_1.

Perform the same tests with another coil connected to T_8. Mark these leads T_5 and T_2. Perform the same test with the last coil connected to T_9 to identify the T_3 and T_6 leads.

Connect L_1 to T_1, L_2 to T_2, L_3 to T_3, and T_4 to T_7, T_5 to T_8, T_6 to T_9, and operate the motor. The motor should operate in the same direction as before and operate quietly.

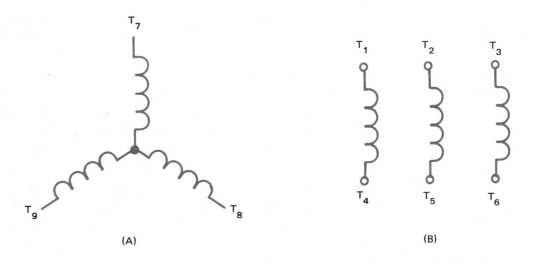

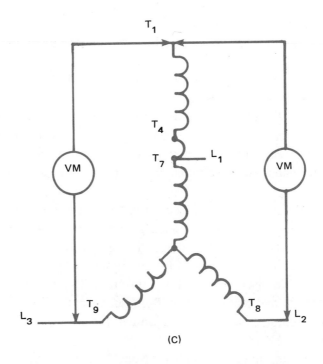

Fig. 14–22 Star or Wye connected motor; A) internal Star point lead marking; B) coil group lead marking C) Checking for proper coil lead markings on Wye connected, dual-voltage motor

Delta Connected Motors

Another connection pattern for three-phase motors is the Delta connected motor. It is so named because the resulting schematic pattern looks like the Greek letter Delta (Δ).

A method of identifying and connecting these leads is necessary because it is different than the Star or Wye connected motor.

Properly connecting the leads of a *Delta* connected, three-phase, dual-voltage motor presents a problem if the lead markings are destroyed.

First, the electrician must determine if the motor is Delta connected or Star connected. Both motors have nine leads if they are dual-voltage motors. However, the Delta-connected motor has three sets of three leads that have continuity and the Star-connected motor has only one set of three.

To proceed, a sensitive ohmmeter is needed to find the middle of each group of three leads. The ohm values are low when using the dc power of an ohmmeter, so use care in identifying the center of each coil group. Label the center of each group T_1, T_2, and T_3 respectively. Using masking tape, temporarily label the other leads of the T_1 group as T_4 and T_9. See Figure 14–23 A.

Temporarily mark the ends of the T_2 group as T_5 and T_7, and mark the ends of the T_3 group as T_6 and T_8.

Connect the lower motor voltage rating using lines 1, 2, and 3 to T_1, T_4, and T_9. The other coils will have induced voltage, so be careful not to touch the other loose leads to each other or to you!

Disconnect the power and connect the lead marked T_4 to T_7. Reconnect the power as before and read the voltage between T_1 and T_2. If the markings are correct the voltage should be about twice the applied line voltage. If it reads about 1.5 times the line voltage, reconnect T_4 to the lead marked T_5. If the voltage T_1 to T_2 then goes to 220, reconnect T_9 to T_7, thereby reversing both coils. When the voltage T_1 to T_2 equals twice the applied line voltage, mark the leads connected together as T_4 from the T_1 group connected to T_7 of the T_2 group.

Now use the third coil group. Leave the lower line voltage connected to the first group as before. Test and connect the leads so that when T_9 is connected to a lead of the third group, the T_1-to-T_2 voltage is twice the applied line voltage. Mark the lead connected to T_9 as T_6 and the other end of the coil group as T_8.

To double check, disconnect the line's lead from T_9 and reconnect to T_7; disconnect the line lead from T_1 and reconnect it to T_2; disconnect the line lead from T_4 and reconnect it to T_5. The motor should run in the same direction as before. If it does not, recheck the lead markings.

To check further, move the line leads from T_7 to T_8, from T_5 to T_6, and from T_2 to T_3. Start the motor. Rotation should be the same as in the previous steps. *Be careful! Voltage is induced into other windings.* (See Figure 14–24).

MOTOR NAMEPLATES

Motor nameplates provide information vital to the proper selection and installation of the motor. Most useful data given on the nameplate refers to the electrical characteris-

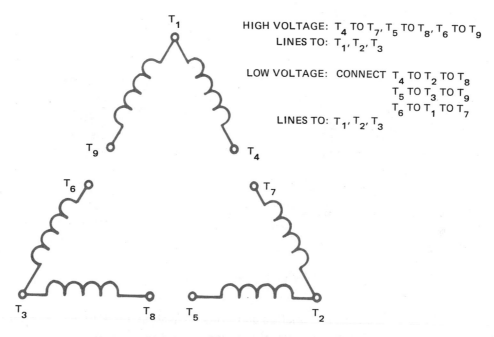

HIGH VOLTAGE: T_4 TO T_7, T_5 TO T_8, T_6 TO T_9
LINES TO: T_1, T_2, T_3

LOW VOLTAGE: CONNECT T_4 TO T_2 TO T_8
T_5 TO T_3 TO T_9
T_6 TO T_1 TO T_7
LINES TO: T_1, T_2, T_3

Fig. 14–23   Nine leads of a Delta connected, three-phase, dual-voltage motor

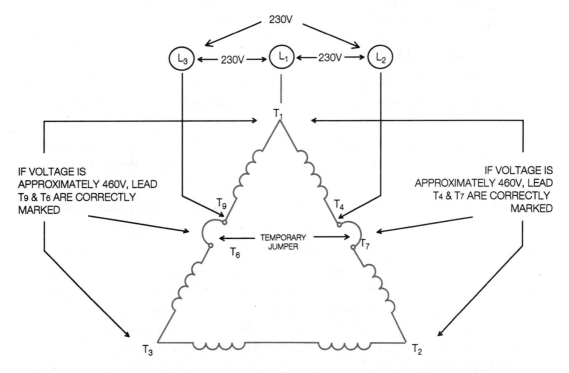

Fig. 14–24   Illustration of voltage tests used to determine correct lead markings on a Delta motor

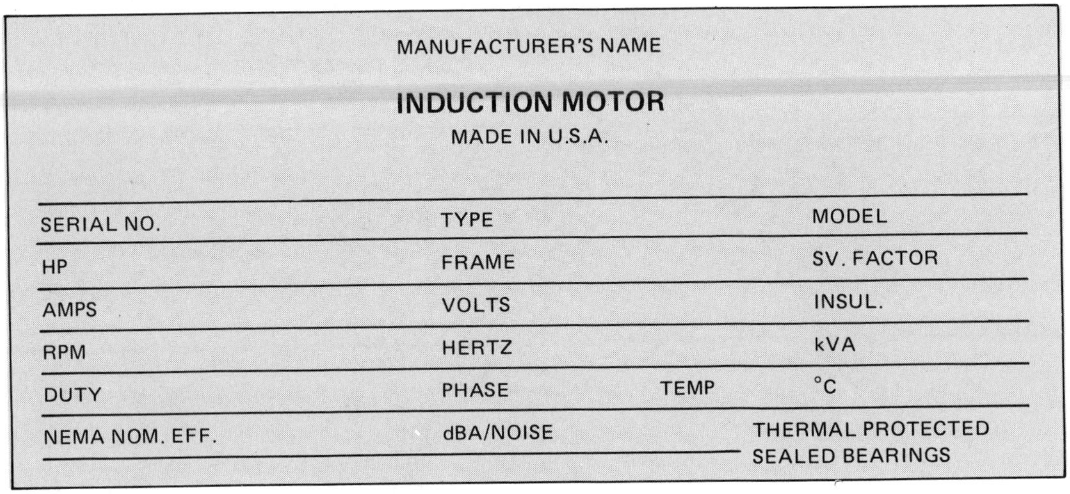

MANUFACTURER'S NAME			
INDUCTION MOTOR			
MADE IN U.S.A.			
SERIAL NO.	TYPE		MODEL
HP	FRAME		SV. FACTOR
AMPS	VOLTS		INSUL.
RPM	HERTZ		kVA
DUTY	PHASE	TEMP	°C
NEMA NOM. EFF.	dBA/NOISE		THERMAL PROTECTED SEALED BEARINGS

Fig. 14–25   Typical motor nameplate

tics of the motor. Given this information and using the National Electrical Code, the electrician can determine the conduit, wire, and starting and running protection sizes. (The NEC gives minimum requirements.)

The design and performance data given on the nameplate is useful to maintenance personnel. The information is vital for the fast and proper replacement of the motor, if necessary. For a better understanding of the motor, typical information found on motor nameplates is described as follows (Figure 14–25).

- The manufacturer's name
- *Type* identifies the type of the enclosure. This is the manufacturer's coded identification system.
- *Serial number* is the specific motor identification. This is the individual number assigned to the motor, similar to a social security number for a person. It is kept on file by the manufacturer.
- The *model number* is an additional manufacturer's identification, commonly used for ordering purposes.
- *Frame size* identifies the measurements of the motor.
- *Service factor (or SF)–a* service factor of 1.0 means the motor should not be expected to deliver more than its rated horsepower. The motor will operate safely if it is run at the rated horsepower times the service factor, maximum. Common service- factors are 1.0 to 1.15. It is recommended that the motor *not* be run continuously in the service factor range. This may shorten the life expectancy of the insulation system.

- *Amperes* means the current drawn from the line when the motor is operating at rated voltage and frequency at the fully rated nameplate horsepower.

- *Volts* should be the value measured at the motor terminals and should be the value for which the motor is designed.

- The *class of insulation* refers to the insulating material used in winding the motor stator. For example, in a Class B system, the maximum operating temperature is 130°C; for Class F, it is 155°C; and for Class H, it is 180°C.

- *RPM* (or *r/min*) means the speed in revolutions per minute when all other nameplate conditions are met.

- *Hertz* is the frequency of the power system for which the motor is designed. Performance will be altered if it is operated at other frequencies.

- *Duty* is the cycle of operation that the motor can safely operate. "Continuous" means that the motor can operate fully loaded 24 hours a day, If "intermediate" is shown, a time interval will also appear. This means the motor can operate at full load for the specified period. The motor should then be stopped and allowed to cool before starting again.

- *Ambient temperature* specifies the maximum surrounding air temperature at which the motor can operate to deliver the rated horsepower.

- *Phase* entry indicates the number of voltage phases at which the motor is designed to operate.

- *kVA* is a code letter which indicates the locked rotor kVA per horsepower. This is used to determine starting equipment and protection for the motor. A code letter table is found in the National Electrical Code.

- *Efficiency* is expressed in percent. This value is found on standard motors as well as "premium efficiency" motors.

- *Noise* – some motors are designed for low noise emission. The noise level given on the nameplate is measured in "dBA" units of sound.

- *Manufacturer's notes* – list specific features of the motors, such as "thermal protected" and/or "scaled bearings."

ALTITUDE

Manufacturers' guarantees for standard motor ratings are usually based on operation at any altitude up to 3,300 feet. Motors suitable for operation at an altitude higher than 3,300 feet above sea level are of special design and/or have a different insulation class. For example, standard motors having a service factor of 1.15 may be operated up to an altitude of 9,900 feet by utilizing the service factor. At an altitude of 9,900 feet, the service factor would be 1.00. It may be necessary to de-rate the motor or use a larger frame size.

SUMMARY

Three-phase induction motors use a squirrel-cage winding in the rotor. There are no electrical connections to the rotor, but current is induced into the rotor windings by electromagnetic induction. The squirrel-cage winding produces a magnetic field that is pushed and pulled by the stator magnetic field.

The rotor is supported by a steel shaft that must rotate. The shaft is allowed to rotate with the application of different types of bearings and various lubrications. Synchronous speed, speed regulation and percent slip are all calculations used in determining the speed of the rotor. Motor electrical characteristics such as power factor and starting current are related to the electrical design of the motor.

If the motor lead markings become destroyed, the leads can be re-marked according to the procedures outlined in this unit. Motor nameplate data is critical information to be used when ordering replacement motors. Some nameplate information is essential for proper replacement of the operating characteristics and other data is used to size the electrical supply and the motor protection.

ACHIEVEMENT REVIEW

A. Answer the following statements and questions.

1. List the essential parts of a squirrel-cage induction motor. _____

_____

2. State two advantages of using a squirrel-cage induction motor. _____

_____

3. State two disadvantages of a squirrel-cage induction motor. _____

_____

_____

4. List the two factors which determine the synchronous speed of an induction motor.

_____

_____

_____

5. Explain how to reverse the direction of rotation of a three-phase, squirrel-cage induction motor._____

_____

_____

_____

6. A four-pole, 60-hertz, three-phase, squirrel-cage induction motor has a full-load speed of 1,725 r/min. Determine the synchronous speed of this motor.

    _____

    _____

    _____

7. What is the percent slip of the motor given in question 6? _____

8. Why is the term squirrel-cage applied to this type of three-phase induction motor?

    _____

    _____

    _____

B. Select the correct answer for each of the following statements and place the corresponding letter in the space provided.

9. Who or what determines if large induction motors may be started at full voltage across the line? _____
    a. maximum motor size
    b. rated voltage
    c. the power company
    d. department of building and safety

10. The power factor of a three-phase, squirrel-cage induction motor operating unloaded is _____
    a. the same as with full load.
    b. very poor.
    c. very good.
    d. average.

11. The power factor of a three-phase, squirrel-cage induction motor operating with full load _____
    a. improves from no load.
    b. decreases from no load.
    c. remains the same as at no load.
    d. becomes 100 percent.

12. The squirrel-cage induction motor is popular because of its characteristics of _____
    a. high percent slip.
    b. low percent slip.
    c. simple, rugged construction.
    d. good speed regulation.

13. The speed of a squirrel-cage induction motor depends on _____
    a. voltage applied.
    b. frequency and number of poles.
    c. field strength.
    d. current strength.

14. Speed is calculated using the formula _____

    a. $p = \dfrac{120 \times f}{r/min}$ c. $RPM = \dfrac{p \times f}{120}$

    b. $RPM = \dfrac{120 \times p}{f}$ d. $RPM = \dfrac{120 \times f}{p}$

C. Draw the following connection diagrams.

15. Show the connection arrangement for the nine terminal leads of a wye-connected three-phase motor rated at 230/460 volts for operation at 460 volts, three phase.

16. Show the connection arrangement for the nine terminal leads of a wye-connected three-phase motor rated at 230/460 volts for operation at 230 volts, three-phase.

U • N • I • T

15

STARTING THREE-PHASE, SQUIRREL-CAGE INDUCTION MOTORS

OBJECTIVES

After studying this unit, the student will be able to

- state the purpose of an across-the line magnetic starting switch.
- describe the basic construction and operation of an across-the-line starter.
- state the ratings for the maximum sizes of fuses required to provide starting protection for motors in the various code marking groups.
- describe what is meant by running overload protection.
- draw a diagram of the connections for an across-the-line magnetic starter with reversing capability.

Alternating-current motors do not require the elaborate starting equipment that must be used with direct-current motors. Most three-phase, squirrel-cage induction motors with ratings up to 10 horsepower are connected directly across the full line voltage. In some cases, motors with ratings greater than 10 horsepower also can be connected directly across the full line voltage. Across-the-line starting usually is accomplished using a magnetic starting switch controlled from a pushbutton station.

The electrician regularly is called upon to install and maintain magnetic motor starters. As a result, the electrician must be very familiar with the connections, operation, and troubleshooting of these starters. The National Electrical Code (NEC) provides information on starting and running overload protection for squirrel-cage induction motors. A comprehensive study of motor controls is given in the Delmar text (*ELECTRIC MOTOR AND MOTOR CONTROL*).

ACROSS-THE-LINE MAGNETIC STARTER

In the simplest starting arrangement, the three-phase, squirrel-cage motor is connected across full line voltage for operation in one direction of rotation. The magnetic switch used for starting, has three heavy contacts, one auxiliary contact, three motor overload relays, and an operating coil. The magnetic switch is called a *motor starter* if it has overload protection. Older motor starters already in service may have used two overload relays. Three overload relays are now required by the National Electrical Code in new installations.

The wiring diagram for a typical across-the-line magnetic starter is shown in figure 15–1A. The three heavy contacts are in the three line leads feeding the motor. The auxiliary contact acts as a sealing circuit around the normally open start pushbutton when the motor is operating. As a result, the relay remains energized after the start button is released. The four contacts of the across-the-line magnetic starter are operated by the magnetic starter coil controlled from a pushbutton station, as shown in figure 15–1B.

Figure 15–2A shows a typical pushbutton station. Two pushbuttons are housed in a pressed steel box. The start pushbutton is normally open and the stop pushbutton is normally closed, as shown in the diagram (figure 15–2B).

STARTING PROTECTION (BRANCH-CIRCUIT PROTECTION)

In figure 15–1A, a motor-rated disconnect switch is installed ahead of the magnetic starter. The safety switch is a three-pole, single-throw enclosed switch. It has a quick-break spring action and is operated externally. The motor circuit switch contains three cartridge fuses which serve as the starting protection for the motor. These fuses must have sufficient capacity to handle the starting surge of current to the motor. The fuses protect the installation from possible damage resulting from defective wiring or faults in the motor windings. This combination may be available in one enclosure (figure 15–3). (See NEC *Article 430.*)

Briefly, the National Electrical Code gives the following information on starting protection for squirrel-cage induction motors.

1. The maximum size fuses permitted to protect motors are rated at 300 percent of the full-load current of the motor for nontime-delay fuses, and 175 percent for time-delay fuses.

NOTE: If the required fuse size as determined by applying the given percentages does not correspond with the standard sizes of fuses available, and if the specified overcurrent protection is not sufficient to handle the starting current of the motor, then the next higher standard fuse size may be used. In no case can the fuse size exceed 400 percent of the full-load current of the motor for nontime-delay fuses and 225 percent of the full-load current for time-delay fuses. (See the National Electrical Code.)

The marking system for squirrel-cage induction motors was developed by the National Electrical Manufacturers Association (NEMA). Note that the fuses used to protect motors with different code letter identifications varies from 150 percent to 300 per-

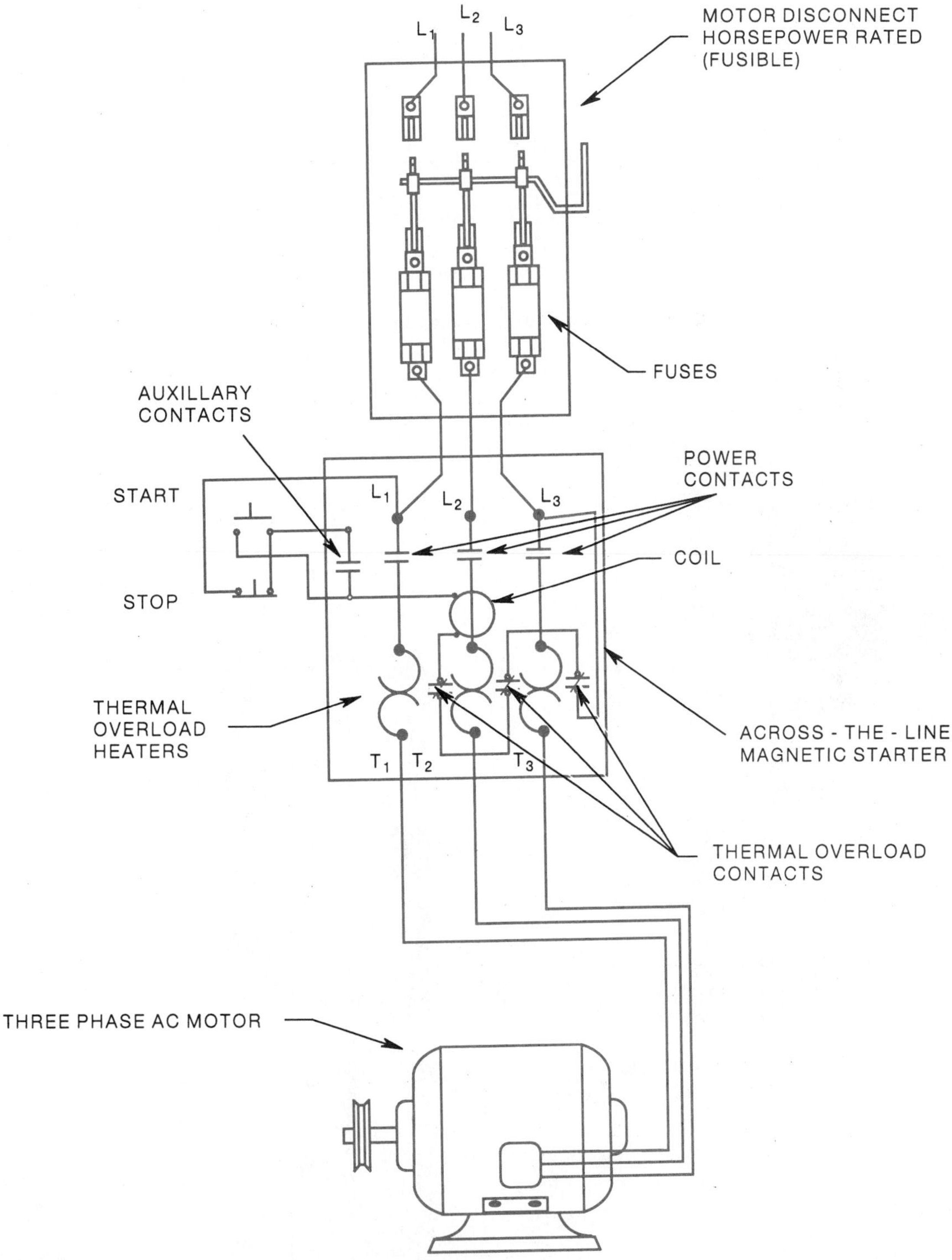

Fig. 15–1A   A wiring diagram for an across-the-fine magnetic starter

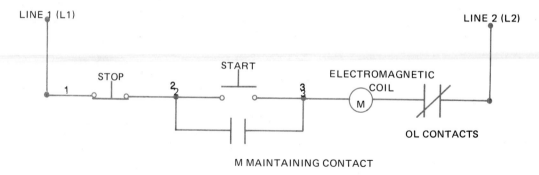

Fig. 15–1B   Elementary diagram of the control circuit for the starter

Fig. 15–2A   Start-stop general purpose control station (*Photo courtesy of Square D Company*)

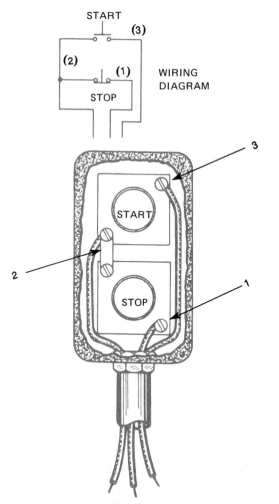

Fig. 15–2B   A pushbutton station and wiring diagram

Fig. 15–3   Combination starter with fusible disconnect switch
(*Photo courtesy of Square D Company*)

cent of the rated full-load current, NEC *Table 430-152*. The difference is in the starting current surges and is due to differences in the design and construction of the rotor.

Rotors are constructed with different characteristics. Figure 15–4 shows the various types of rotor construction and the associated code letter. The applications of motors with these code letters is also indicated. The design of the rotor affects the amount of current needed to produce the rotor magnetic field. Code letter A has high starting torque and relatively low starting current. The code book chart 430–7(b) will indicate that a code letter A motor will have less locked rotor kVA than other motors. This calculation indicates there is less starting current for the same voltage for a code motor. The chart in figure 15–4 gives some broad categories of motors. A, B to E, F to V.

An ac magnetic starter is shown in figure 15–5. The starter consists of power contacts that are used to open and close the circuit to the motor. As ac is applied to the magnetic coil, the magnet draws the contacts closed and connects the line power to the motor power. In addition to connecting the line power, the magnetic starter has an add-on block at the bottom to provide for running overload protection. See unit 16 for a detailed operation of the magnetic starter.

Example 1.   A three-phase, squirrel-cage induction motor with a nameplate marking of code letter F is rated at 5 hp, 230 volts. According to the National Electrical Code, this motor has a full-load current per terminal of 15.2 amperes. The starting protection shall not exceed 300 percent of the rated current for squirrel-cage motors with nontime-delay fuses. Thus, the starting protection is 15.2 × 3 = 45.6 amperes.

INDUCTION MOTOR WITH CODE LETTER A

THIS TYPE OF MOTOR HAS A HIGH-RESISTANCE ROTOR WITH SMALL ROTOR BARS NEAR THE ROTOR SURFACE. THIS MOTOR HAS A HIGH STARTING TORQUE AND LOW STARTING CURRENT.

APPLICATIONS:

METAL SHEARS, PUNCH PRESSES, AND METAL DRAWING MACHINERY.

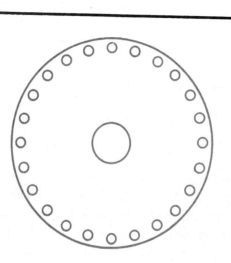

INDUCTION MOTOR WITH CODE LETTERS B-E

THIS TYPE OF MOTOR HAS A HIGH-REACTANCE AND LOW-RESISTANCE ROTOR. THIS MOTOR HAS A RELATIVELY LOW STARTING CURRENT AND ONLY FAIR STARTING TORQUE. IT HAS LARGER CONDUCTORS DEEP IN THE ROTOR IRON.

APPLICATIONS:

MOTOR-GENERATOR SETS, FANS, BLOWERS, CENTRIFUGAL PUMPS, OR ANY APPLICATION WHERE A HIGH STARTING TORQUE IS NOT REQUIRED.

INDUCTION MOTOR WITH CODE LETTERS F-V

THIS TYPE OF MOTOR HAS A RELATIVELY LOW-RESISTANCE AND LOW-INDUCTIVE REACTANCE ROTOR. THIS MOTOR HAS A HIGH STARTING CURRENT AND ONLY FAIR STARTING TORQUE. IT HAS LARGE CONDUCTORS NEAR THE ROTOR SURFACE.

APPLICATIONS:

MOTOR-GENERATOR SETS, FANS, BLOWERS, CENTRIFUGAL PUMPS, OR ANY APPLICATION WHERE A HIGH STARTING TORQUE IS NOT REQUIRED.

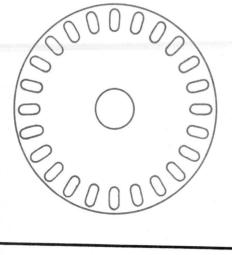

Fig. 15-4 Various types of rotor laminations

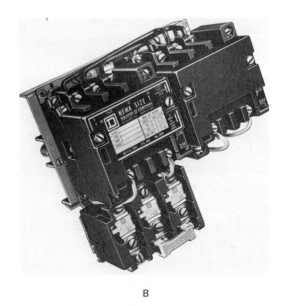

B

Fig. 15–5 A) A magnetic starter includes the contactor and the overload section (*Courtesy of Allen-Bradley*) B) Ac reversing magnetic motor starter. The elementary diagram of the starter is shown in figure 15–8) (*Photo courtesy of Square D Company*)

Since a 45.6-ampere fuse cannot be obtained (see NEC *Section 240-6)*, the next larger size of fuse (50 amperes) should be used. For motor branch-circuit protection, the motor current listed in the appropriate table of the National Electrical Code should be used. The full-load current, as stated on the motor nameplate, is not used for this purpose.

RUNNING OVERLOAD PROTECTION

Many motor starters installed in the United States use a thermal-type overload assembly. The assembly is normally located beneath the contactor and is directly attached to the magnetic contactor. The overload monitoring system is designed to measure the amount of current flowing to the motor through the contactor. This is done by connecting thermal sensors called *heaters* in series with the motor current. The heaters are sized to produce a certain amount of heat with a specified current through them. They are calibrated to cause a thermally-operated switch to open when there is sustained heat. The heat is caused by too much current flow to the motor which indicates the motor is jammed or is working too hard and is overloaded. The thermal sensors are varied as seen in figure 15–6. The heater sensors with the associated trip-overload relays are pictured. The National Electrical Code requires the use of three thermal overload units as running overload protection. Although new installations require three overload relays, the electrician will work on many older installations which have only two overload relays. These were installed before the three overload relay requirement became effective. The overload relay unit may be either three individual units, or a common block containing the three heaters and only one trip switch contact unit reacting from any one of the heaters.

These overload heater units are made of a special alloy. Motor current through these units causes heat to be generated. In one type, a small bimetallic strip is located next to each of the two heater units. When an overload on a motor continues for a period of approximately one to two minutes, the excessive heat developed by the heater units causes the bimetallic strips to expand. As each bimetallic strip expands, it causes the normally closed contacts in the control circuit to open. The main relay coil is deenergized and disconnects the motor by opening the main and auxiliary contacts. Melting alloy overloads (solder pots) also are commonly used. The heat generated by the overload melts the solder pot to release a ratchet which trips the control circuit contacts. Many motor starters are provided with electronic overload relays. The sensors are actually current transformers that measure the exact current flowing to the motor and will provide a trip signal to the magnetic starter if the current is too high for too long.

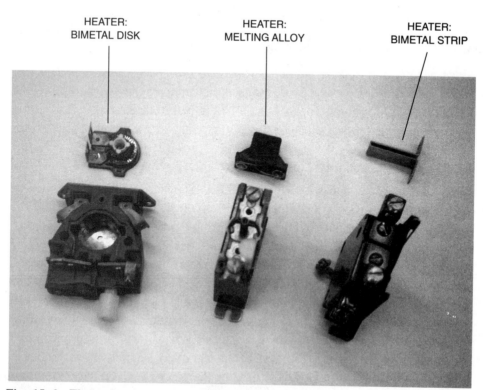

HEATER:
BIMETAL DISK

HEATER:
MELTING ALLOY

HEATER:
BIMETAL STRIP

Fig. 15–6   Thermal overload relays. Shown are the bimetal disk, the melting alloy style and the bimetal strip (From Keljik, *Electric Motors and Motor Controls*, copyright 1995 by Delmar Publishers)

Before the motor can be restarted at the pushbutton station, the overload contacts in the control circuit *must be allowed to cool* before being reclosed (reset). When the reset button in the magnetic starter is pressed, the overload contacts in the control circuit are reset to their normally closed position. The motor then can be controlled from the push-button station.

The National Electrical Code requires that the running overload protection in each phase be rated at not more than 125 percent of the full-load current rating for motors which are marked with a temperature rise of not more than 40 degrees Celsius (see NEC *Article 430, Part C).*

Example 2. Using the motor full-load current rating from the nameplate data, determine the running overcurrent protection for a three-phase, 5-hp, 230-volt squirrel-cage induction motor with a rated full-load current of 14.5 amperes and a temperature rise of 40 degrees Celsius. The running overcurrent protection is $14.5 \times 1.25 = 18.1$ amperes.

For this motor, heater overload units rated to trip at 18.5 amperes are required for the magnetic starter. Where the overload relay so selected is not sufficient to start this motor, the next higher size overload relay is permitted, but not to exceed 140 percent of the motor full-load current rating. Actual motor nameplate currents are used to establish the overload protection.

AUXILIARY CONTACTS

In addition to the standard contacts, a starter may be provided with externally attached auxiliary contacts, sometimes called *electrical interlocks* (figure 15–7). These auxiliary contacts can be used in addition to the holding circuit contacts, and the main or power contacts which carry the motor current. Auxiliary contacts are rated to carry only control circuit currents of 0-15 amperes, not motor currents. Versions are available with either normally open or normally closed contacts. Among a wide variety of applications, auxiliary contacts are used to

- control other magnetic devices where sequence operation is desired.
- electrically prevent another controller from becoming energized at the same time (such as reverse starting), called interlocking.
- make and break circuits of indicating or alarm devices, such as pilot lights, bells, or other signals.

Auxiliary contacts are packaged in kit form, and can be added easily in the field.

ACROSS-THE-LINE MOTOR STARTER WITH REVERSING CAPABILITY

The direction of rotation of a squirrel-cage induction motor must be reversed for some industrial applications. To reverse the direction of rotation of 3 phase motors interchange any two of the three line leads.

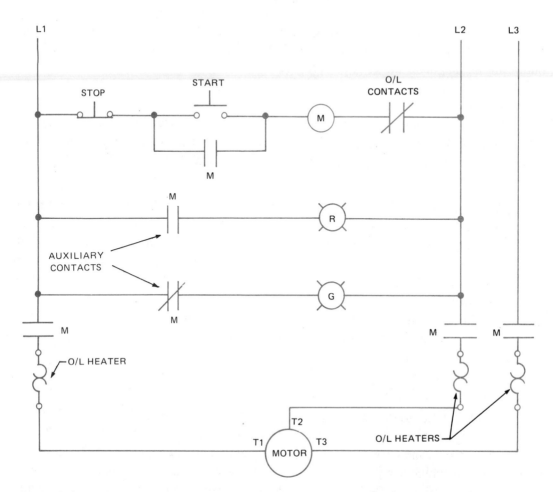

Fig. 15–7   Electrical interlocks (auxiliary contacts) switch pilot lights in this circuit

Figure 15–8 is an elementary wiring diagram of a motor starter having a reversing capability. When the three power reverse contacts are closed, the phase sequence at the motor terminals is different from that when the three power forward contacts are closed. Two of the line leads feeding to the motor are interchanged when the three reverse power contacts close.

The control circuit has a pushbutton station with Forward, Reverse, and Stop pushbuttons. The control circuit requires a mechanical and an electrical interlocking system provided by the push buttons. Electrical interlocking means that if one of the devices in the control circuit is energized, the circuit to a second device is open and cannot be closed until the first device is disconnected. Mechanical interlocks, shown by the broken lines in figure 15–8, are used between the forward and reverse coils and pushbuttons.

Note in figure 15–8 that when the forward pushbutton is pressed, it breaks contact with terminals 4 and 5, opening the reverse coil circuit, and makes contact between termi-

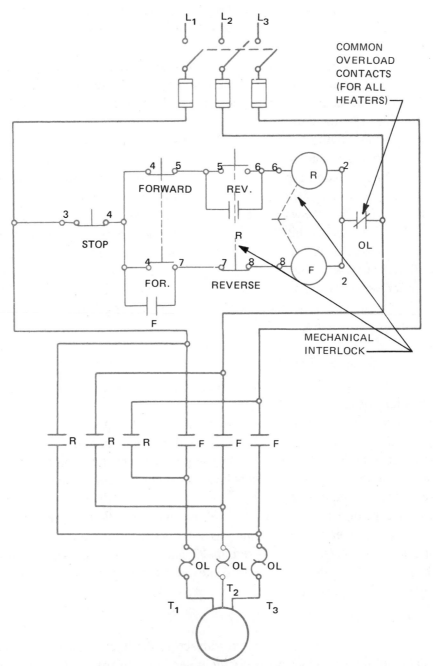

Fig. 15–8   Elementary diagram of an across-the-line magnetic starter with reversing capabilities

nals 4 and 7. As a result, coil F is energized and the forward contacts close. The motor now rotates in the forward direction. If the reverse pushbutton is pressed, it will break contact between terminals 7 and 8, and open the circuit to coil F. This causes all forward contacts to open. As the reverse pushbutton is depressed farther, it closes the contact between terminals 5 and 6 and energizes coil R. All reverse contacts are now closed and the motor rotates in the reverse direction. If the stop button is pressed, the contact between terminals 3 and 4 is opened, the control circuit is interrupted, and the motor is disconnected from the three-phase source. The National Electrical Code requirements for starting and running overload protection which apply to the across-the-line motor starter also apply to this type of motor starter.

Figure 15–8 and figure 15–9 are actually the same motor controller. Figure 15–8 is drawn in an elementary diagram. It has the control circuit in a schematic style, which shows the electrical relationship of the components. It shows the power contact of the magnetic starter below the schematic, and the electrical relationship of the motor control components. Figure 15–9 shows the same components, but in the approximate physical location of the components. This style of drawing is called a *wiring diagram*. Many electricians find it is easier to wire a panel from the wiring diagram as it shows physical location as well as general wire routing. Many electricians find it easier to troubleshoot from a schematic, or elementary diagram, as it shows electrical sequence of operation more clearly. It is important that you know how to read both types of drawing and be able to transfer from one to the other.

DRUM REVERSING SWITCH

A drum reversing switch (figure 15–10A) may be used to reverse the direction of rotation of squirrel-cage induction motors.

The motor is started in the forward direction by moving the handle of the drum reversing switch from the off position to the forward (F) position. The connections for this drum controller in both the forward and reverse positions are shown in figure 15–11. In the forward position, the switch connects line 1 to motor terminal 1, line 2 to motor terminal 2, and line 3 to motor terminal 3.

To reverse the direction of rotation, the drum switch handle is moved to the reverse (R) position. In the reverse position, line 1 is still connected to motor terminal 1. However, line 2 is now connected to motor terminal 3, and line 3 is connected to motor terminal 2. When the handle of the drum switch is moved to the off position, all three line leads are disconnected from the motor.

SUMMARY

Many squirrel-cage motors are started with across-the-line motor starters. The motor and branch circuit should include short-circuit protection such as fuses or circuit breakers. The motor must also have running-overload protection. This protection is usually found with the starter and is in the form of thermal-overload heaters and the associated overload relay. The overload relay is designed to open the control circuit to the motor starter in the event of a sustained overload on the motor. Motors can be automatically controlled

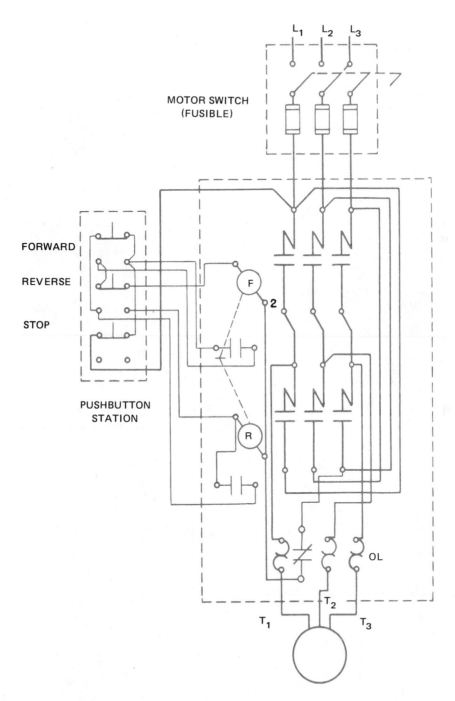

Fig. 15–9 A panel or wiring diagram of an across-the-line magnetic starter with reversing capability

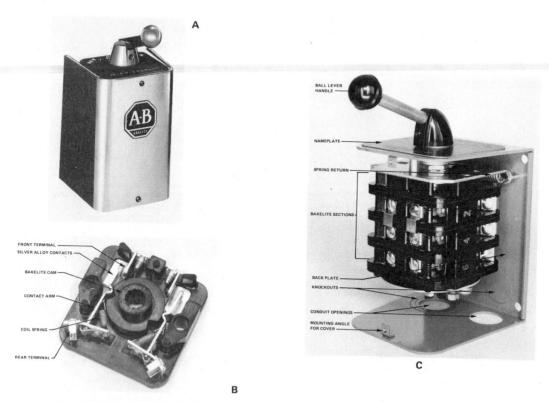

Fig. 15–10   A) Reversing drum switch B) A bakelite section of a drum switch
C) Bakelite section with cover removed (*Photos courtesy of Allen-Bradley Company*)

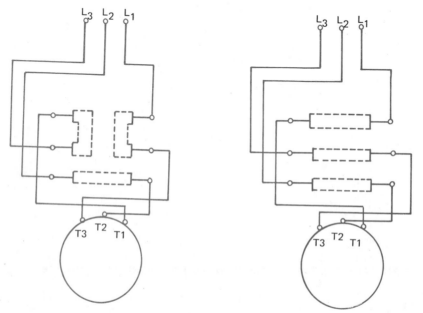

Fig. 15–11   Connections for a drum reversing switch. Left, reverse- right, forward

through the use of a magnetic starter or may be manually controlled thorough the use of a drum-type controller. In either case, a three-phase motor may be reversed by interchanging two of the three-line connections to the motor.

ACHIEVEMENT REVIEW

1. What is the purpose of starting protection for a three-phase motor? _____

    _____

    _____

    _____

2. What is the purpose of running overload protection for a three-phase motor?

    _____

    _____

    _____

3. What is meant by the code letter markings of squirrel-cage induction motors?

    _____

    _____

    _____

4. List some of the industrial applications for squirrel-cage induction motors with code letter classification A. _____

    _____

5. List some of the industrial applications for squirrel-cage induction motors with code letter classifications B to E. _____

    _____

    _____

    _____

6. List some of the industrial applications for squirrel-cage induction motors with code letter classifications F to V. _____

    _____

    _____

    _____

7. A three-phase motor (code letter J) has a full-load current rating of 40 amperes, and a temperature rise of 40°C.

    a. What are the maximum size fuses that can be used for branch-circuit protection?

    _____

    _____

b. What size heaters would be used for running overcurrent protection?

_____

_____

8. What is the maximum starting protection allowed by the National Electrical Code?

_____

_____

_____

16

CONTROLLERS FOR
THREE-PHASE MOTORS

OBJECTIVES

After studying this unit, the student will be able to

- describe the basic sequence of actions of the following types of controllers when used to control three-phase ac induction motors:

    jogging-type controller,

    quick-stop ac controller (plugging),

    dynamic braking controller,

    resistance starter controller,

    automatic autotransformer compensator,

    automatic controller for wound-rotor induction motors,

    wye-delta controller, and

    automatic controller for synchronous motors.

- identify and use the various National Electrical Code sections pertaining to controllers and remote control circuits for motors.
- state why ac adjustable speed drives are used.
- list the types of adjustable speed drives.
- describe the operating principles of different ac adjustable speed drives.
- list the advantages and disadvantages of some units.

The industrial electrician is required to install, maintain, and repair automatic ac controllers which start up and provide speed control for squirrel-cage induction motors, wound-rotor induction motors, and synchronous motors.

MOTOR CONTROLLERS WITH JOGGING CAPABILITY

Many industrial processes require that the driven machines involved in the process be inched or moved small distances. Motor controllers designed to provide control for this type of operation are called jogging controllers. *Jogging* is defined as the quickly repeated closure of a controller circuit to start a motor from rest for the purpose of accomplishing small movements of the driven machine.

An across-the-line magnetic motor switch may be used to provide jogging control if the proper type of pushbutton station is used in the control circuit. Such a pushbutton station is called a *start-jog-stop station*.

Figure 16–1 A is a diagram of the connections for a three-phase, squirrel-cage induction motor connected to a jogging-type, across-the-line motor starting switch.

Figure 16–1B shows the starter with the cover removed. Note in figure 16–1 that the connections and operation of the start and stop pushbuttons are the same as those of a standard pushbutton station with start and stop positions. The connections for the jog pushbutton, however, are more complex and should be studied in detail. When the jog pushbutton is pressed, coil M is energized, main contacts M close, and the motor starts turning. The small auxiliary contacts M also close, but do not function as a sealing circuit around the jog pushbutton because pushing the jog pushbutton also opens the sealing circuit. As a result, as soon as the jog pushbutton is released, coil M is deenergized and all M contacts open. Before the jog pushbutton returns to its normal position, the sealing contacts M open and thus the control circuit remains open. This control also can be used for standard start-stop operations. In summary then, repeated closures of the jog pushbutton start the motor momentarily so that the driven machine can be inched or jogged to the desired position.

QUICK-STOP AC CONTROLLER (PLUGGING)

Some industrial applications require that three-phase motors be stopped quickly. If any two of the line leads feeding a three-phase motor are reversed, a counter torque is set up which brings the motor to a quick standstill before it begins to rotate in the reverse direction. If the circuit is interrupted at the instant the motor begins to turn in the opposite direction, the rotor will just stop. This method of bringing a motor to a quick stop is called *plugging*. The motor controller required to provide this type of operation is an across-the-line magnetic motor starter with reversing control and a special plugging relay. The plugging relay is belt-driven from an auxiliary pulley on the motor shaft, or onto a through shaft motor.

The connections for a quick-stop ac controller are shown in figure 16–2A. The controller itself is shown in figure 16–2B. When the start pushbutton is pressed, relay coil F is energized. As a result, the small, normally closed F contacts open. These contacts are connected in series with the reverse coil, which locks out reverse operation. In addition, the other small, normally open F contacts close and maintain the start pushbutton circuit. When the start button is released, the circuit of coil F is maintained through the sealing circuit, main contacts F close, and the rated three-phase voltage is applied to the motor terminals. Then, the motor comes up to speed and contacts PR of the plugging relay close.

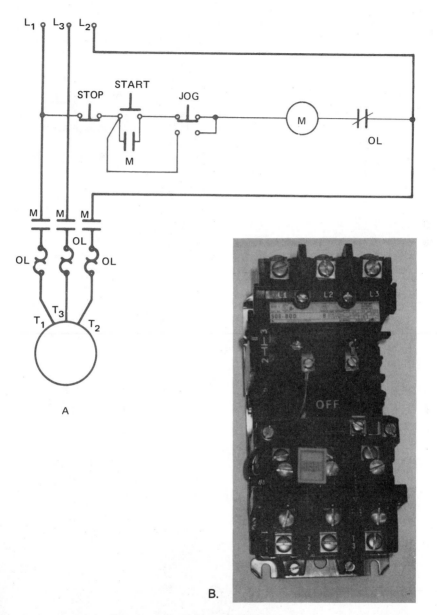

**Fig. 16–1 A) Elementary diagram connections for a three-phase motor with jogging
B) Magnetic starter removed from enclosure**

When the stop pushbutton is pressed, the F relay coil is deenergized. As a result of this action, the main F contacts for the motor circuit open and disconnect the motor from the three-phase source. In addition, the small F sealing contacts open, deenergizing the holding or sealing circuit around the start pushbutton. Finally, the small F contacts in series with the reverse relay close to their normal position and the reverse relay coil is energized.

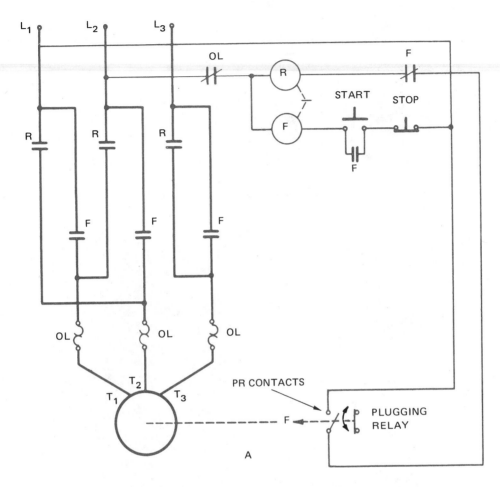

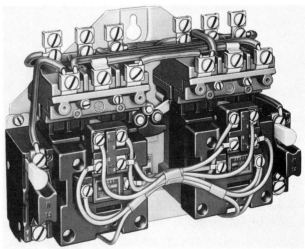

Fig. 16–2   A) Elementary circuit with a plugging relay B) Ac full voltage reversing starter, size 1 (*Photo courtesy of Allen-Bradley Company*)

The main R contacts now close to reconnect the three-phase line voltage to the motor terminals. The connections of two line leads are interchanged. The resulting reversing countertorque developed in the motor brings it to a quick stop. At the moment the motor begins to turn in the reverse direction, contacts PR open due to the mechanical action of the PR relay unit. Coil R is deenergized and the R contacts open and interrupt the power supply to the motor. Since the motor is just beginning to turn in the opposite direction, it comes to a standstill. The motor supplies the mechanical power to drive a disk which causes contacts PR to close when the motor is in operation.

DYNAMIC BRAKING WITH INDUCTION MOTORS

It should be recalled from the study of dc controllers that dynamic braking is a method used to help bring a motor to a quicker stop without the extensive use of friction brakes. In this application, dynamic braking means that the motor involved is used as a generator. An energy dissipating resistance is connected across the terminals of the motor after it is disconnected from the line.

Dynamic braking also can be applied to induction motors. When the stop pushbutton is pressed, the motor is disconnected from the three-phase source and the stator windings are excited by a dc source. A stationary magnetic field is developed by the direct current in the stator windings. As the squirrel-cage rotor revolves through this stationary field, a high rotor current is created. This rotor current reacts with the stationary field of the stator to produce a countertorque that slows and stops the motor.

Figure 16–3 is a diagram of an ac motor installation with an across-the-line magnetic motor starter and dynamic braking capability.

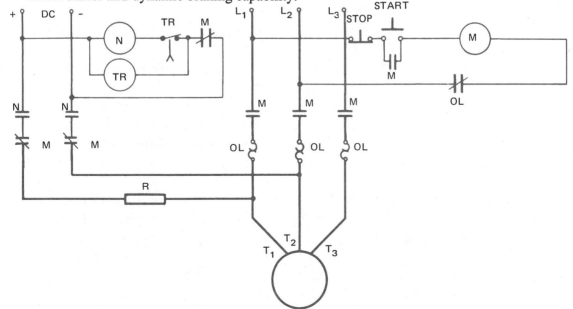

Fig. 16–3   A circuit for dynamic braking of an ac motor

When the start pushbutton is pressed, coil M is energized. At this instant, the main M contacts close and connect the motor terminals to the three-phase source, and the auxiliary, normally open M contacts close and provide a maintaining circuit around the start pushbutton. When relay coil M is energized, the normally closed contacts M in the dc control circuit open, with the result that both the main dc relay coil N and the time-delay relay coil TR are deenergized and interlocks in the dc circuit open. The three-phase voltage applied to the motor terminals causes the motor to accelerate to the rated speed.

When the stop pushbutton is pressed, coil M is deenergized. At this moment, a number of actions occur: 1) the main contacts M open and disconnect the motor from the three-phase source; 2) the auxiliary M contacts open (these contacts act as a maintaining circuit); 3) protective interlocks M in the dc circuit close; and 4) the auxiliary, normally closed contacts M in the dc control circuit close and energize the time-delay relay and the main relay coil N. Energizing relay coil N causes the closing of contacts N so that dc voltage is connected on the ac windings through a current-limiting resistance. As a result, the motor comes to a quick stop. Following a definite period after the motor has stopped, measured in seconds, relay coil TR operates and opens contacts TR to cause coil N to become deenergized. Thus, contacts N open and disconnect the motor windings from the dc source. The controller now is ready for the next starting cycle.

Timing contacts are shown in their deenergized condition. Timers are either *on-delay* or *off-delay* and are used in motor control work. The actual timer mechanism varies depending on the vintage of the controller and the manufacturer. See unit 13 for complete description of the timer symbols and operations.

RESISTANCE STARTER CONTROLLER

When a squirrel-cage induction motor is connected directly across the rated line voltage, the starting current may be 300 percent to 600 percent of the rated current of the motor. In large motors, this high current may cause serious voltage regulation problems and overloading of industrial power feeders.

The starting current of a squirrel-cage induction motor can be reduced by using a resistance starter controller. This type of controller inserts equal resistance values in each line wire at the instant the motor is started. After the motor accelerates to a value near its rated speed, the resistance is cut out of the circuit and full line voltage is applied to the motor terminals.

Figure 16–4A is a diagram of the circuit connections for a resistance starter. A photo of this starter is figure 16–4B. When the start button is pressed, main relay coil M is energized. The main contacts close and connect the motor to the three-phase source through the three resistors (R). The circuit for coil M is maintained through the small auxiliary contacts (3 and 4) which act as a sealing circuit around the normally open start pushbutton. When the main contacts of relay coil M are closed, a mechanical device, called a *definite time relay,* is started. After a predetermined time elapses, the definite time contacts

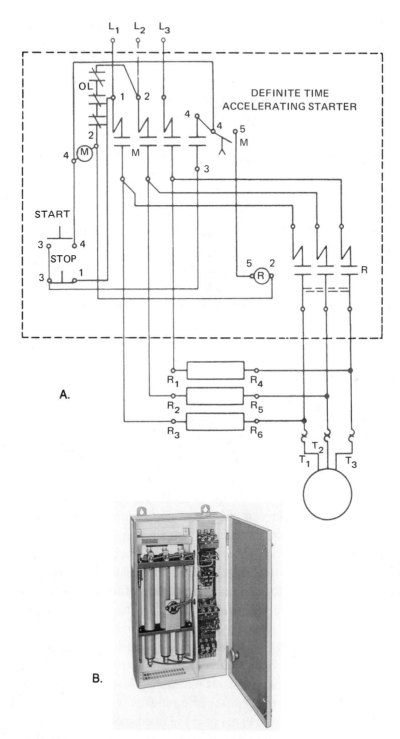

A.

B.

Fig. 16–4 **A) Definite time acceleration with resistance starting B) Two-point primary resistance starter rated at 25 hp, 600 V** (*Photo courtesy of Allen-Bradley Company*)

close and energize coil R. Coil R causes three sets of contacts to close and shunt out the three resistors. Thus, the motor is connected directly across the rated line voltage with no interruption of the power line (closed transition).

When the stop button is pressed, the circuits of both coil M and coil R are opened. This causes the opening of the main contacts, the sealing contacts, and the contacts which shunt the series resistors. As a result, the motor is disconnected from the three-phase source.

The starting current in the resistance starter causes a relatively high voltage drop in the three resistors. Because of this, the voltage across the motor terminals at start is low. As the motor accelerates, the current decreases, the voltage drop across the three resistors decreases, and the terminal voltage of the motor increases gradually. A smooth acceleration is obtained because of this gradual increase in the terminal voltage. However, it may be unwise to select resistance starting for many starting tasks because of the energy dissipated in the starting resistors.

The National Electrical Code provides guidelines on the selection of the correct fuse sizes for starting protection on a branch motor circuit containing a squirrel-cage induction motor with a resistance starter. In addition, the Code specifies the running overload protection required, and the wire sizes required on the branch motor circuit. (See NEC *Article 430*.)

AUTOMATIC AUTOTRANSFORMER COMPENSATOR

The automatic autotransformer (figure 16–5A) compensator basically operates in the same manner as a manual starting compensator. A manual compensator is used to manually connect an autotransformer in series with the motor during the starting cycle. A *compensator* is another name for an autotransformer starter. By inserting an autotransformer, the voltage to the motor is reduced and therefore the starting current is reduced. The starting current drawn from the line is reduced. Because the autotransformer steps down the voltage to the motor, the secondary (motor) current is higher than the primary current. The result is that the current drawn from the line is much less than if the motor were started with a reduced voltage through a resistor. The automatic compensator has the advantage in that it can be pushbutton-controlled from a convenient location. Figure 16–5B is a typical schematic wiring diagram for an automatic autotransformer compensator.

If the start button is pressed, a circuit is established from line 1 through the following devices to line 2: the normally closed stop button, the start button, contacts TR (timing relay) to relay coil S (start coil), and the normally closed overload contacts. Coil M is energized, closing the normally open small contacts M to provide the maintaining circuit.

When coil S is energized, all contacts marked S close. The three autotransformers are connected in wye across the three-phase line and supply reduced voltage to the motor. The motor begins to accelerate to a value near the rated speed.

As shown in figure 16–5B, a second circuit is established through the timer relay by contacts TR. The timer relay begins operating as soon as it is energized. After a definite

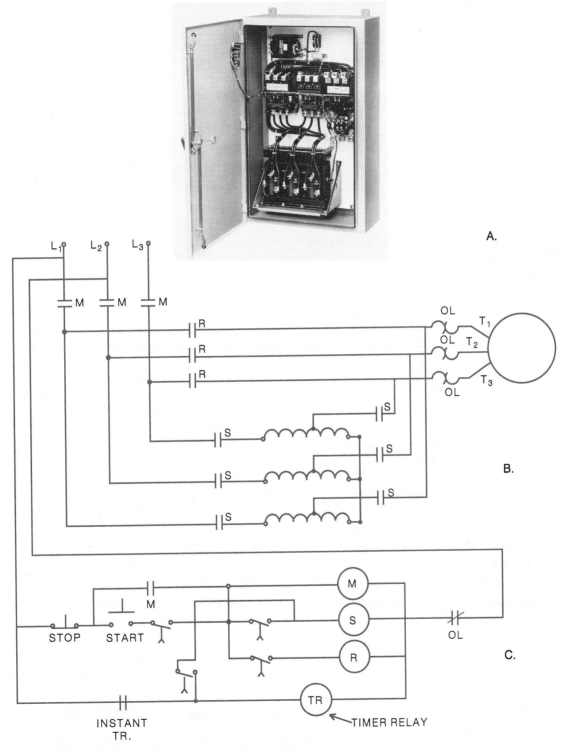

Fig. 16–5 A) Reduced voltage starter, autotransformer type, with a pneumatic timer (*Courtesy of Allen-Bradley Company*) B) Elementary diagram of an autotransformer compensator for starting an induction motor

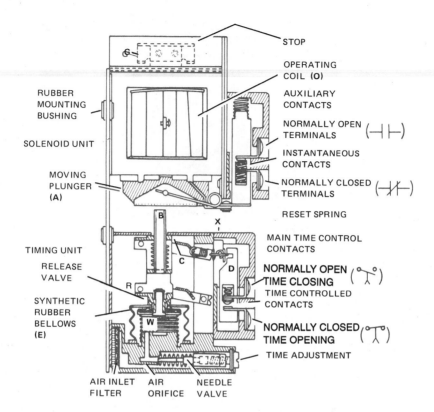

STOP

OPERATING
COIL (O)

AUXILIARY
CONTACTS

NORMALLY OPEN
TERMINALS (⊣ ⊢)

INSTANTANEOUS
CONTACTS

NORMALLY CLOSED (⊣/⊢)
TERMINALS

RESET SPRING

MAIN TIME CONTROL
CONTACTS

NORMALLY OPEN
TIME CLOSING

TIME CONTROLLED
CONTACTS

NORMALLY CLOSED
TIME OPENING

TIME ADJUSTMENT

RUBBER
MOUNTING
BUSHING

SOLENOID UNIT

MOVING
PLUNGER
(A)

TIMING UNIT

RELEASE
VALVE

SYNTHETIC
RUBBER
BELLOWS
(E)

AIR INLET   AIR   NEEDLE
FILTER   ORIFICE   VALVE

Fig. 16–6 Cross section of an ac on-delay timer, which provides time delay after the coil is energized. It is shown with the coil energized and the timer timed out. Schematic wiring symbols are shown (de-energized positions) for various portions of the timer.

timed period (figure 16–6) the timer relay mechanically actuates all TR contacts. Coil S is deenergized as a result. This coil was used to start the three-phase motor on reduced voltages. At the same time, coil R (run coil) is energized and closes the running contacts to apply the rated three-phase voltage to the motor terminals.

At this time, the motor is operating on full voltage. If the stop button is pressed, the holding coil circuit for coils M and R opens. As a result, the R contacts to the motor open and the motor stops. When coil R is deenergized, the normally closed R contacts close the pilot motor circuit. The pilot motor runs to set all TR contacts for the next starting cycle.

The National Electrical Code rulings for starting and running protection also apply to motors operated with either a manually operated starting compensator or an automatic autotransformer compensator.

AUTOMATIC CONTROLLER FOR WOUND-ROTOR INDUCTION MOTORS

Manual speed controllers, such as the faceplate type or the drum type, may be used to provide speed control for wound-rotor induction motors in industrial applications. If the resistance in the rotor circuit of a wound-rotor induction motor is to be used only on

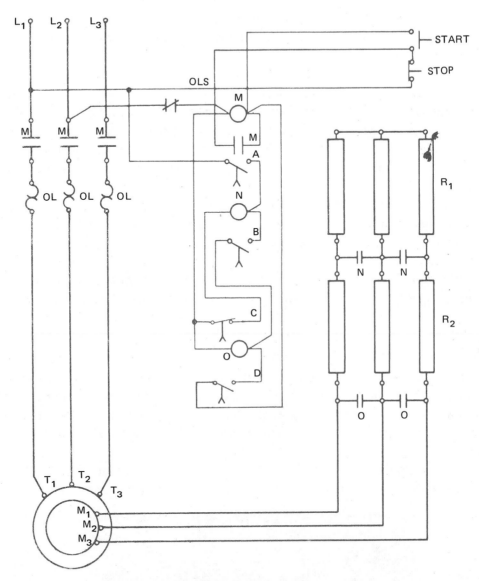

Fig. 16–7 Elementary diagram of an automobile controller for a wound-rotor induction motor

starting, then an automatic controller may be used, figure 16–7. In this case, resistors in the rotor circuit are automatically removed by contactors arranged to operate in sequence at definite time intervals.

As shown in figure 16–7, when the start button is pressed, the main relay coil M is energized. The main contacts are closed to connect the stator circuit of the motor directly across the three-phase line voltage. All of the resistance of the controller is inserted in the secondary circuit of the motor as it begins to accelerate.

After the start button is released to its normally open position, the small auxiliary contacts M act as a maintaining circuit to keep the circuit of coil M closed. Contacts A are held open for a timed period (seconds) by a mechanical or electronic device (figure 16–8A, B, and C). When the A contacts close, coil N is energized through the normally closed contacts C, and all N contacts close to shunt out the R_1 resistors in the rotor circuit.

Contacts B also are held open for a definite number of seconds by a mechanical or electronic device. When B contacts close, coil O is energized, all O contacts close, and all resistance is cut out of the rotor circuit. At the same time, the C contacts open and deenergize coil N which then opens contacts N. The D contacts then close and maintain a closed circuit through coil O.

When the stop pushbutton is pressed, relay coil M is deenergized, and contacts M open to disconnect the motor from the line. Coil O also is deenergized and contacts O open, with the result that all of the resistance is inserted in the rotor circuit for the next starting cycle.

The National Electrical Code regulations for wire size, starting overload protection, and running overload protection also apply to both manual speed controllers and automatic controllers used with wound-rotor induction motors.

WYE-DELTA CONTROLLER

Figure 16–9A shows a simple method by which a three-phase, delta-connected motor can be started on reduced voltage by connecting the stator windings of the motor in wye during the starting period. Figure 16–9B shows the actual starter. After the motor accelerates, the windings are reconnected in delta and placed directly across the rated three-phase voltage.

When the start button is pressed, the main M contacts close, and relay coil Y and time-delay relay TR are energized. Coil Y causes contacts Y to close and the windings of

A.

Fig. 16–8   A) Solid-state timing relays with different plug-in program keys
(*Courtesy of Square D Company*)

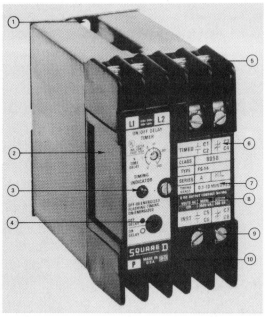

1. STANDARD INDUSTRIAL CONTROL RELAY MOUNTING
2. REMOVABLE TIMER COVER PROTECTS TIME DELAY AND MODE SETTING
3. LED (LIGHT EMITTING DIODE) TIMING INDICATOR
4. CONVERTIBLE TIME DELAY MODE SHOWS THROUGH COVER
5. ONE N.O. AND ONE N.C. TIMED NEMA B150 HARD OUTPUT CONTACTS (10 AMPERE CONTINUOUS)
6. TERMINALS CLEARLY MARKED
7. FIVE TIMING RANGES FROM 0.05 SECONDS TO 10 HOURS
8. MARKING AREA
9. SELF-LIFTING PRESSURE WIRE CONNECTORS
10. OPTIONAL INSTANTANEOUS NEMA B150 HARD OUTPUT CONTACTS (10 AMPERE CONTINUOUS)

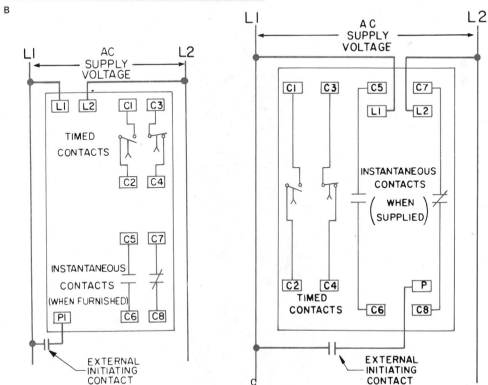

Fig. 16–8   B-C) Solid-state timing relays with different plug-in program keys (*Courtesy of Square D Company*)

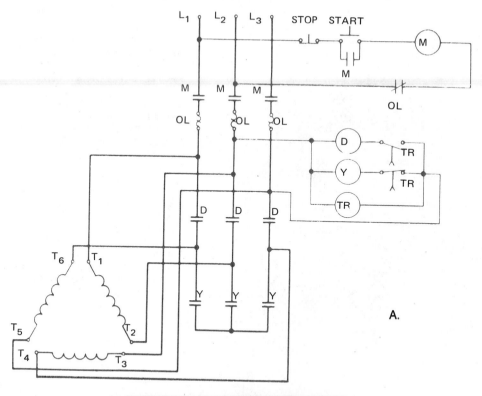

A.

B.

Fig. 16–9 A) Elementary diagram of wye-delta motor starting B) Photo of wye delta, 200 hp, closed transition

the motor are connected in wye. If the line voltage is 230 volts, the voltage across each winding is:

$$\frac{230}{1.73} = 133 \text{ volts}$$

The voltage across each winding is only 58 percent of the line voltage when the windings are connected in wye at the start position. (See 3-phase Voltage in Electricity Three.)

After a definite period of time, the time-delay relay TR opens the circuit of relay coil Y and the Y contacts open.

Then, the time-delay relay TR closes the circuit of relay coil D. All D contacts are closed and the motor winding connections are changed from wye to delta. Full line voltage is applied to the motor windings and the motor operates at its rated speed.

Motors started by a wye-delta controller must have the leads of each phase winding brought out to the terminal connection box of the motor. In addition, the phase windings must be connected in delta for the normal running position. NOTE: The electrician should never attempt to operate a three-phase, wye-connected motor with this type of controller. This is due to the fact that there will be an excessive voltage applied to the motor windings in the run position when the windings are connected in delta by the controller.

AUTOMATIC CONTROLLER FOR SYNCHRONOUS MOTORS

Synchronous motors may be started by means of an across-the-line magnetic motor starting switch, a manual starting compensator, or an automatic starting compensator. Dynamic braking may be provided by the controller.

Figure 16–10 is a diagram of the connections for a synchronous motor controller with dynamic braking. When the start button is pressed, main relay coil M is energized. The four normally open M contacts close and the two normally closed M contacts open. Three-phase voltage is applied to the motor terminals. When the motor accelerates to a speed near the synchronous speed, the dc field circuit is energized by secondary controls.

When the stop button is pressed, main relay coil M is deenergized. The M contacts open and disconnect the motor terminals from the three-phase line. The two normally closed M contacts reconnect the motor windings through the resistors and the dc field remains energized. As a result, the synchronous motor acts as an ac generator and delivers electrical energy to the two R resistors. The use of this type of controller results in a more rapid slowing of a synchronous motor.

The National Electrical Code provides guidelines for branch-circuit fuse protection and running overload protection for branch circuits feeding three-phase synchronous motors, and for allowable conductor sizes for branch circuits feeding synchronous motors. Local building and electrical code authorities should be consulted before installations are made with motors and controllers which do not comply with National Electrical Code rulings.

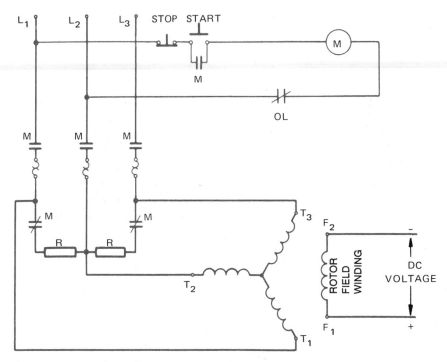

Fig. 16–10   Synchronous motor controller with dynamic braking

SOLID-STATE REDUCED VOLTAGE STARTERS

Solid-state devices and equipment are used for reduced voltage motor starting, electrical energy saving control circuits, variable speed drives, motor protection and other applications. A motor starter consists of a control circuit, a motor power circuit, and protective devices for the wiring and the motor. The functions of a starter are performed by contactors and overload relays in electromechanical motor starters. In solid-state starters, the control functions are performed by semiconductors. They are controlled by integrated circuits and microprocessors to provide the protective functions, operating instructions, and control.

Construction and Operation

The solid-state reduced-voltage starter provides a smooth, stepless acceleration of a three-phase induction motor. This is accomplished by gradually turning on six power SCRs (silicon controlled rectifiers). Two SCRs per phase are connected in a back-to-back or reverse parallel arrangement, figure 16–11. The SCRs are mounted on a heat (dissipating) sink to make up a power pole (phase). Each power pole contains the gate firing circuits as discussed in unit 13. An integrated thermal sensor is also provided to deenergize the starter if an over-temperature condition exists.

The firing circuitry on each power pole is controlled by a logic module. These modules monitor the starter for correct start up and operating conditions. Some motor starters provide a visual indicator of the starting condition through the use of light-emitting diodes (LEDs).

The SCRs are connected back to back so that they may pass ac and control the amount of voltage. The current-limiting starter is a common type; it is designed to maintain the motor current at a constant level throughout the acceleration period. A current-limit potentiometer adjustment is provided to preset this current. A starter with current ramp acceleration is designed to begin acceleration at a low current level and then increase the current during the acceleration period.

As indicated in figure 16–11, this starter includes both *start* and *run* contactors. The start contacts are in series with the SCRs; the run contacts are in parallel with the combination of SCRs and start contacts. When the starter is energized, the start contacts close. The motor acceleration is then controlled by phasing-on the SCRs, When the motor reaches full speed, the run contacts close and the motor is connected directly across the lines (closed transition). At this point, the SCRs are turned off and the start contacts open. Under full speed running conditions, the SCRs are out of the circuit, eliminating SCR power losses during the run cycle. This feature saves energy; it also guards against possible damage due to overvoltage transients. With the starter in the deenergized position, all contacts are open, isolating the motor from the line. This open circuit condition protects against accidental motor rotation as a result of SCR misfiring and/or SCR damage caused by overvoltage transients. A solid-state reduced-voltage starter is shown in figure 16–12. Field connections are very similar to those for electromechanical starters.

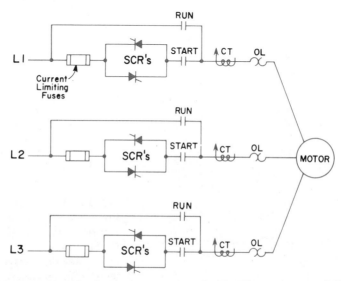

Fig. 16–11   Solid-state reduced voltage starter power circuit (*Photo courtesy of Allen-Bradley*)

Reduced Voltage Operation

To reduce the voltage applied to the motor in a solid-state starter, the SCRs can be turned on by the "gate" electrode for any desired part of each half cycle. Usually the *SCRs* turn off as the current wave reaches zero. They stay off until gated on again in the next half cycle. Some devices can vary the switching and timing. By switching the controlled current gating, the effective ac voltage can be varied to the motor. This voltage can be varied from zero to full voltage as required. The voltage is applied at some preset minimum value that can start the motor rotating. As the motor speed builds up, the SCR "on" time is gradually increased. The voltage is increased until the motor is placed across the line at full voltage. Mechanical shock is reduced and the current inrush can be regulated and controlled as desired (figure 16–13). The solid-state reduced-voltage starter can replace any of the electromechanical starters already described for reduced voltage starting.

CODE REFERENCES FOR MOTOR CONTROLLERS

The following sections of the National Electrical Code are concerned with motor controllers and remote control circuits.

Fig. 16–12 Solid-state reduced voltage starter power circuit (*Photo courtesy of Allen-Bradley Company*)

Fig. 16–13 SCR controller section, the regulating part of the starter. The controller determines to what degree the SCR's should be phased on, thereby controlling the voltage applied to the motor (*Photo courtesy of Allen-Bradley Company*)

1. *Sections 430-8* and *430-9* refer to the identification of motors and controllers with respect to controller nameplate ratings and terminal markings.

2. *Section 430-C* is concerned with overload protection.

3. *Section 430-37* gives the number of running overcurrent relays required for different electrical systems.

4. *Section 430-D* is concerned with branch-circuit protection.

5. *Sections 430-71* to *430-74* are concerned with the control circuits of controllers.

6. *Sections 430-81* to *430-90* are concerned with controller installations.

7. *Sections 430-101* to *430-113* cover motor disconnecting means.

AC ADJUSTABLE SPEED DRIVES

Adjustable speed drives have a flexibility that is particularly useful in specialized applications. For this reason, these drives are widely used throughout industry for conveyors used to move materials, hoists, grinders, mixers, pumps, variable speed fans, saws, and crushers. The advantages of ac drives include the maximum utilization of the driven equipment, better coordination of production processes, and reduced wear on mechanical equipment.

The ac induction motor is the major converter of electrical energy into another usable form. About two-thirds of the electrical energy produced in the United States is delivered to motors.

Much of the power that is consumed by ac motors goes into the operation of fans, blowers, and pumps. It has been estimated that approximately 50% of the motors in use are for these types of loads. Such loads are particularly appropriate to look at for energy savings. Several alternate methods of control for fans and pumps have been developed and show energy savings over traditional methods of control.

Fans and pumps are designed to meet the maximum demand of the system in which they are installed. Often, however, the actual demand varies and may be much less than the design capacity. Such conditions are accommodated by adding outlet dampers to fans or throttling valves to pumps. These controls are effective and simple, but affect the efficiency of the system. Other forms of control have been developed to adapt machinery to varying demands. These controls do not decrease the efficiency of the system as much as the traditional methods of control. One of the newer methods is the direct variable speed control of the fan or pump. This method produces a more efficient means of flow control compared to the other existing methods.

In addition to a tangible reduction in the power required to operate equipment and machinery resulting from the use of adjustable speed drives, other benefits include extended bearing life and pump seal life.

WOUND ROTOR AC MOTORS

Wound rotor motor drives use a specially constructed ac motor to accomplish speed control. The windings of the motor rotor are brought out of the motor through slip rings on the rotor shaft. Figure 16–14 shows an elementary diagram of a wound rotor motor with an adjustable speed drive. These windings are connected to a controller which places variable resistors in series with the windings. The torque performance of the motor can be controlled using these variable resistors.

Wound rotor motors are more common in the larger sizes, in the range of 300 horsepower and above.

Features of Wound Rotor Motors

Wound rotor motors have the following advantages which make them suitable for a variety of applications:

- **Cost** – the initial cost is moderate for the high horsepower units.
- **Control** – not all the power need be controlled, resulting in a moderate size and simple controller.
- **Construction** – the simple construction of the motor and control lends itself to maintenance without the need for a high level of training.
- **High inertia loads** – the drive works well on high inertia loads.

Disadvantages of Wound Rotor Motors

- **Custom motor** – the motor has a rotor wound with wire, slip rings, and is not readily available.
- **Efficiency** – the drive does not maintain a high efficiency at low speeds.
- **Speed range** – the drive usually is limited to a speed range of two to one.

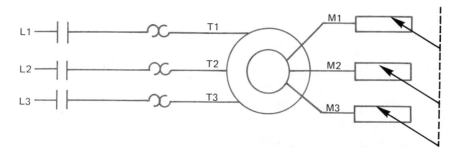

Fig. 16–14 Elemetary diagram of an adjustable speed drive wound rotor motor

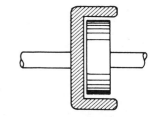

Fig. 16-15   Spider rotor coil magnet rotated within a steel drum (From Alerich, *Electric Motor Control*, 4th edition, copyright 1988 by Delmar Publishers Inc.)

TYPES OF ADJUSTABLE SPEED DRIVES

Several types of variable speed drives can be used with wound rotor induction motors. These drives are eddy current (magnetic) drives, variable pitch drives, and adjustable frequency drives.

Eddy Current (Magnet) Drives

The eddy current drive couples the motor to the load magnetically (figure 16-15). The electromagnetic coupling is a simple way to obtain an adjustable output speed from the constant input speed of squirrel cage motors. There is no mechanical contact between the rotating members of the eddy current drive; thus, there is no wear. Torque is transmitted between the two rotating units by an electromagnetic reaction created by an energized rotating coil winding. The rotation of the ring with relation to the electromagnet generates eddy currents and magnetic fields in the ring. Magnetic interaction between the two units transmits torque from the motor to the load. The slip between the motor and the load can be controlled continuously with great precision.

Torque can be controlled using a thyristor in an ac or dc circuit, or by using a rheostat to control the field through slip rings. When the eddy current drive responds to an input or command voltage, the speed of the driven machine changes. A further refinement can be obtained in automatic control to regulate and maintain the output speed. The magnetic drive can be used with nearly any type of actuating device or transducer that can provide an electrical signal. For example, the input can be provided by static controls and sensors which detect liquid level, air and fluid pressures, temperature, and frequency.

Magnetic eddy current drives are used for applications requiring an adjustable speed such as cranes, hoists, fans, compressors, and pumps (figure 16-16).

Variable Pitch Drives

The speed of an ac squirrel cage induction motor depends upon the frequency (hertz) of the supply current and the number of poles of the motor. The equation expressing this relationship is:

$$RPM = \frac{60 \times Hertz}{Pairs\ of\ Poles}$$

Fig. 16-16 **Two magnetic drives driven by 100-hp induction motors mounted on top** *(Photo courtesy of Electric Machinery Manufacturing Group.* From Alerich, *Electric Motor Control,* 4th Edition, copyright 1988 by Delmar Publishers Inc.)

A frequency changer may be used to vary the speed of this type of motor. A possible method is to drive an alternator through an adjustable mechanical speed drive.

The voltage is regulated automatically during frequency changes. An ac motor drives a variable cone pulley or sheave, which is belted to another variable pulley on the output shaft (figure 16-17). When the relative diameters of the two pulleys are changed, the speed between the input and the output can be controlled. As the alternator speed is varied, the output frequency varies, thereby varying the speed of the motor, or motors, connected electrically to the alternator supply.

Adjustable Frequency AC Drives

Adjustable frequency (static solid-state) drives are also commonly called *inverters.* The power conversion losses are greatly reduced when using these transistor controllers for adjustable speed drives. They are available in a range of horsepowers from fractional to 1,000 hp. Adjustable frequency drives are designed to operate standard ac induction motors. This allows them to be added easily to an existing system. (Figure 16-18).

Where energy saving is a major concern the drives are ideal for pumping and fan applications. They are also used for many process control or machine applications where performance is a major concern. Many adjustable speed precision applications were lim-

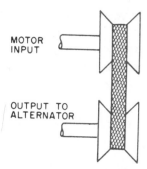

MOTOR
INPUT

OUTPUT TO
ALTERNATOR

Fig. 16-17 **Variable pitch pulley method of obtaining continuously adjustable speed from constant speed shaft** (From Alerich, *Electric Motor Control,* 4th Edition, copyright 1988 by Delmar Publishers Inc.)

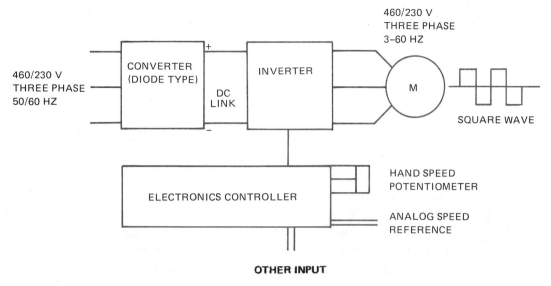

Fig. 16–18   Controller operation

ited to the use of dc motors. By using adjustable frequency controllers with optional dynamic braking, standard squirrel cage motors can now be used in these applications. Municipal, industrial, commercial and mining applications include: sewage, waste water, slurry and booster pumps, ventilation and variable air volume fans; conveyors; production machines and compressors.

AC VARIABLE SPEED/VARIABLE FREQUENCY CONTROLS

AC motors are designed to run at a specific full load speed. This design speed takes into account various losses in the motor including copper losses of the stator and the rotor and other losses such as iron losses, friction, and windage. The end result is some speed less than synchronous speed, which is calculated by the following formula:

$$\text{Synchronous RPM} = \frac{120 \text{ Freq}}{\text{\# poles}}$$

Use this formula to determine the synchronous speed of a motor if the number of poles and the applied frequency are known. The number of poles is usually fixed and the frequency of a normal power feed in the U.S.A. is 60 Hz. Therefore, to operate an ac motor at other than its design speed, either the number of poles or the frequency must be altered. Some *adjustable* speed motors are able to reconnect the poles or connect a separate winding of poles to establish other set speeds. To vary the speed of an ac motor over a wide range of speeds, the applied frequency is altered.

There are two basic techniques for altering the frequency of the applied power through electronic means. Both techniques use the principle of rectifying the three-phase

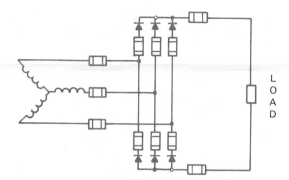

Fig. 16–19   Three-phase, full-wave rectifier with connected load

ac 60 Hz input power to a dc supply (see figure 16–19). Then filter the dc to provide smooth dc to the inverter section of the electronic controller. The amplitude of the output voltage must change with the frequency, because at low frequency the impedance of the motor is low and the voltage must also be reduced to prevent overheating of the motor. Conversely, as the frequency is raised above 60 Hz, the motor's impedance is increased and voltage must also increase to maintain motor torque.

One method of speed control is the variable voltage inverter. The dc voltage applied to the inverter is adjusted, then the pulse is modified to create various frequencies (see figure 16–18).

The other and more common method is known as pulse width modulation, or PWM. The voltage output of the dc-to-ac inverter is really a series of pulses of DC that is stepped to produce a "staircase" approximation of the sine wave at the frequency desired, to control the motor speed. The controller uses a sensor and a set point to determine output frequency and the desired set speed.

The control modules adjust the output of a dc voltage control module as it adjusts the output of the frequency control module. The frequency control module typically drives the output power controller, there are two of these for each of the three phases; one controls the positive half cycles of one phase and one controls the negative half cycles of the same phase. Six output modules are required for three phases.

When connected to a motor the stepped output wave form appears close to a sine wave because of the motor's inductance. Frequency then determines the speed of the motor in conjunction with the set number of poles.

INTRODUCTION TO PROGRAMMABLE CONTROLLERS

Modern motor control requires more exact timing and faster, more consistent operations than those provided by the older electromechanical relays and timers. Because modern manufacturing requirments also rely on flexibility of systems, a fast change system was needed that didn't require extensive redesign, building, handwiring, and testing to create a new motor control scheme.

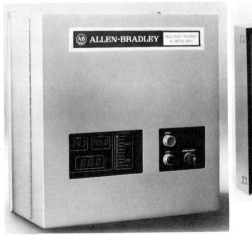
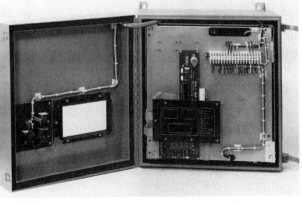

Fig. 16–20   Adjustable frequency motor drive controller (*Photos courtesy of Allen-Bradley Company, Drives Division*)

Along with this change in needs, the microprocessor was improved to make it more reliable in industrial environments and more trouble-free in operation. The combination of these events produced the Programmable Controller, or PC.

Various versions of the PC are made by different manufacturers and have various capabilities. However, all programmable controllers have at least three basic components: The first is a processor, which is a microprocessor that contains the instructions and makes the decisions; the second component is an input/output section, which receives input information from the process it is controlling, and connects it to the controller in the proper format. It also takes the output from the controller and interfaces to the real world of control devices; the third main component is the programmer. This is the operator's access to the controller. The programs are written in ladder logic style, which is familiar to most electricians. This familiarity was the factor that allowed the programmable controller to be so widely accepted as a control system. Controls were easily converted and adapted by current electricians without having to learn extensive new microprocessor programming language.

PCs are used where the control system is likely to be changed frequently and usually where there are multiple functions for the controller to monitor, compare, count, time, or operate. These conditions make the use of a PC economical and practical.

Large control systems may have various input/output cards that must be coordinated according to the "field wiring." For example: ac and dc voltage at various levels, Transistor-Transistor Logic (TTL), Input/Output (IO), Analog I/O, Thermocouple or Binary Coded Decimal (BCD). Figure 16–21 is a programmable controller.

The intent of this book is not to teach programming of the PC, but to familiarize the electrician with the possibilty of motor control using the PC. *A Technician's Guide to Programmable Controllers,* by Richard A. Cox, Delmar Publishers Inc., is an excellent reference to gain further "generic" information.

SUMMARY

There are many methods used to start, stop, jog, and reverse three-phase motors. The basic operations generally use a magnetic controller to supply power to the motor. The reversing controller is used in the plugging operation to momentarily reverse the power and therefore bring the motor to a quick stop. Dynamic braking can also be used to stop an ac motor by applying dc to the motor field. Various methods of reducing the starting

Fig. 16–21   Programmable controller, with one input/output card removed

current to the motor employ the application of reduced voltage at the motor terminals during the starting period. Resistance or reactance can be inserted in series with the motor to reduce the voltage. An autotransformer may be used to reduce the applied voltage. Wound-rotor motors use secondary resistors to keep the starting current to a minimum. Wye-delta starters can be used to reduce the starting current to the motor by changing the configuration of the connections. Solid state starters are now being installed to reduce the inrush current and to control the speed and the stopping characteristics.

Variable-frequency drives use electronics to control the frequency to the motor and therefore control the speed of the motor. Many motor-control schemes can be developed with the use of a programmable controller. This electronic equivalent of a relay system is used to provide exact timing and complex, but changeable, control through the use of a microprocessor-based system.

ACHIEVEMENT REVIEW

1. What is meant by the term jogging? _____

   _____

   _____

2. What is meant by the term plugging? _____

   _____

   _____

3. How is dynamic braking applied to an ac induction motor? _____

   _____

   _____

4. How is dynamic braking applied to a synchronous motor? _____

   _____

   _____

5. Draw a schematic diagram of the connections for the control circuit of an across-the-line magnetic motor switch with jogging capability. Include the main relay coil, the pushbutton station with start, jog, and stop pushbuttons, and the sealing contactor.

6.   What identifying information should appear on a motor controller so that the controller complies with the requirements of the National Electrical Code?

_____

_____

7.   Draw a schematic diagram of an automatic controller used for a three-phase, wound-rotor induction motor.

8. A three-phase, squirrel-cage induction motor has the following ratings: 15 horse-power, 230 volts, 42 amperes per terminal, 40 degrees Celsius, and code classification F. In the spaces in the following table, insert the correct values for fuse protection and running overcurrent protection for this motor when used with each of the types of controllers listed.

Type of Controller	Fuse Protection		Running Overcurrent Protection
	Nontime-Delay	Time-Delay (Dual Element)	
a. Resistance starter			
b. Automatic autotransformer compensator			
c. Across-the-line magnetic motor starting switch with jogging capability			
d. Across-the-line magnetic motor starting switch with plugging capability			

9. What is the purpose of an automatic autotransformer starting compensator?

_____

_____

_____

10. What is the purpose of an automatic controller used with wound-rotor induction motors?_____

_____

_____

11. A wye-delta controller starts the motor at _____
    a. 173 percent of the line voltage.
    b. 58 percent of the line voltage.
    c. full line voltage.
    d. 25 percent of the line voltage.

12. A three-phase, wye-connected motor _____
    a. should never be started by a wye-delta controller.
    b. should always be started by a wye-delta controller.
    c. can be started by a wye-delta controller if proper timers are used.
    d. can be started by a wye-delta controller if proper pushbuttons are used.

13. How are SCRs connected to pass and control ac?_____

_____

14. If a solid-state controller has contactors in the power circuit, in what position are the contacts of start and run in the off position? Why?_____

_____

_____

_____

15. Why are adjustable speed drives used? _____

_____

_____

_____

_____

16. List the types of ac adjustable speed drives. _____

_____

_____

_____

_____

17. How is the speed of the wound rotor motor adjusted? _____

_____

_____

_____

_____

18. How is the eddy current drive coupled to the load? _____

_____

_____

_____

_____

19. How is the ac frequency varied in the mechanical method drive? _____

_____

_____

Unit 16 Controllers for Three-Phase Motors   175

20. What is the formula for calculating ac motor speed?_____

_____

_____

_____

21. What basic devices are provided in the adjustable frequency drive? _____

_____

_____

_____

22. With an apparent high degree of skill required to maintain an adjustable frequency drive control, how does the plant electrician repair one? _____

_____

_____

_____

_____

17

THREE-PHASE, WOUND-ROTOR INDUCTION MOTOR

OBJECTIVES

After studying this unit, the student will be able to

- list the main components of a wound-rotor, polyphase induction motor.

- describe how the synchronous speed is developed in this type of motor.

- describe how a speed controller connected to the brushes of the motor provides a variable speed range for the motor.

- state how the torque, speed regulation, and operating efficiency of the motor are affected by the speed controller.

- demonstrate how to reverse the direction of rotation of a wound-rotor induction motor.

Until the last several years, ac variable speed control was very difficult with a standard motor. Therefore, a different type of motor and control system was developed and used extensively for years. Maintenance electricians must be familiar with this type of motor and control system.

Many industrial motor applications require three-phase motors with variable speed control. The squirrel-cage induction motor cannot be used for variable speed work since its speed is essentially constant. Another type of induction motor was developed for variable speed applications. This motor is called the *wound-rotor induction motor* or *slip-ring ac motor*.

CONSTRUCTION DETAILS

A three-phase, wound-rotor induction motor consists of a stator core with a three-phase winding, a wound rotor with slip rings, brushes and brush holders, and two end shields to house the bearings that support the rotor shaft.

Figures 17–1, 17–2, 17–3, and 17–4 show the basic parts of a three-phase, wound-rotor induction motor.

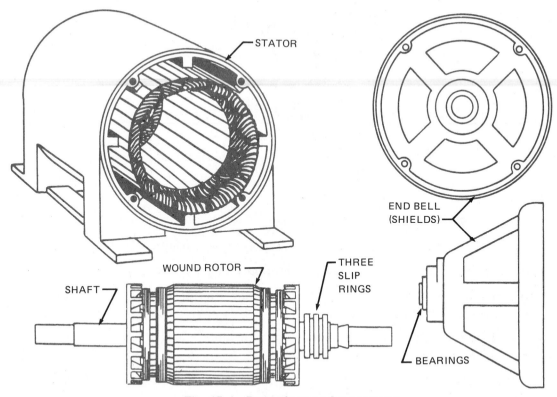

Fig. 17–1   Parts of a wound-rotor motor

Fig. 17–2   Wound stator for a polyphase induction motor (*Photo courtesy of General Electric Company*)

Fig. 17–3   Wound rotor for a polyphase induction motor (*Photo courtesy of General Electric Company*)

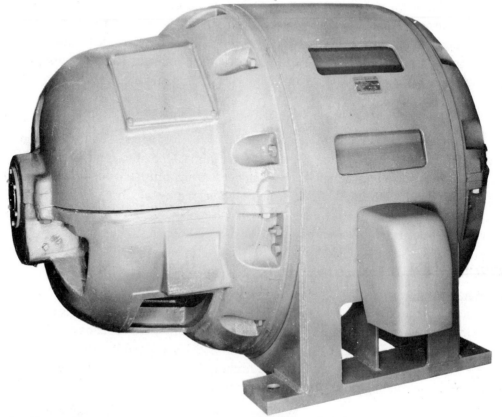

Fig. 17–4   Sleeve bearing, wound-rotor polyphase induction motor (*Photo courtesy of General Electric Company*)

The Stator

A typical stator contains a three-phase winding held in place in the slots of a laminated steel core, figure 17–2. The winding consists of formed coils arranged and connected so that there are three single-phase windings spaced 120 electrical degrees apart. The separate single-phase windings are connected either in wye or delta. Three line leads are brought out to a terminal box mounted on the frame of the motor. This is the same construction as the squirrel-cage motor stator.

The Rotor

The rotor consists of a cylindrical core composed of steel laminations. Slots cut into the cylindrical core hold the formed coils of wire for the rotor winding.

The rotor winding consists of three single-phase windings spaced 120 electrical degrees apart. The single-phase windings are connected either in wye or delta. (The rotor winding must have the same number of poles as the stator winding.) The three leads from the three-phase rotor winding terminate at three slip rings mounted on the rotor shaft. Leads from carbon brushes which ride on these slip rings are connected to an external speed controller to vary the rotor resistance for speed control.

The brushes are held securely to the slip rings of the wound rotor by adjustable springs mounted in the brush holders. The brush holders are fixed in one position. For this type of motor, it is not necessary to shift the brush position as is sometimes required in direct-current generator and motor work.

The Motor Frame

The motor frame is made of cast steel. The stator core is pressed directly into the frame. Two end shields are bolted to the cast steel frame. One of the end shields is larger than the other because it must house the brush holders and brushes which ride on the slip rings of the wound rotor. In addition, it often contains removable inspection covers.

The bearing arrangement is the same as that used in squirrel-cage induction motors. Either sleeve bearings or ball-bearing units are used in the end shields.

PRINCIPLE OF OPERATION

When three currents, 120 electrical degrees apart, pass through the three single-phase windings in the slots of the stator core, a rotating magnetic field is developed. This field travels around the stator. The speed of the rotating field depends on the number of stator poles and the frequency of the power source. This speed is called the synchronous speed. It is determined by applying the formula which was used to find the synchronous speed of the rotating field of squirrel-cage induction motors.

$$\text{Synchronous speed in RPM} = \frac{120 \times \text{frequency in hertz}}{\text{number of poles}} \quad \text{or} \quad S = \frac{120 \times F}{P}$$

$$S = \frac{120 \times f}{p}$$

As the rotating field travels at synchronous speed, it cuts the three-phase winding of the rotor and induces voltages in this winding. The rotor winding is connected to the three slip rings mounted on the rotor shaft. The brushes riding on the slip rings connect to an external wye-connected group of resistors (speed controller), figure 17–5. The induced voltages in the rotor windings set up currents which follow a closed path from the rotor winding to the wye-connected speed controller. The rotor currents create a magnetic field in the rotor core based on transformer action. This rotor field reacts with the stator field to develop the torque which causes the rotor to turn. The speed controller is sometimes called the *secondary resistance control.*

Starting Theory of Wound-Rotor Induction Motors

To start the motor, all of the resistance of the wye-connected speed controller is inserted in the rotor circuit. The stator circuit is energized from the three-phase line. The voltage induced in the rotor develops currents in the rotor circuit. The rotor currents, however, are limited in value by the resistance of the speed controller. As a result, the stator current also is limited in value. In other words, to minimize the starting surge of current to a wound-rotor induction motor, insert the full resistance of the speed controller in the rotor circuit. The starting torque is affected by the resistance inserted in the rotor secondary. With resistance in the secondary, the power factor of the rotor is high or close to unity. This means that the rotor current is nearly in phase with the rotor-induced voltage. If the rotor current is in phase with the rotor-induced voltage, then the rotor magnetic poles are being produced at the same time as the stator poles. This creates a strong magnetic effect, which creates a strong starting torque. As the motor accelerates, steps of resistance in the wye-connected speed controller can be cut out of the rotor circuit until the motor accelerates to its rated speed.

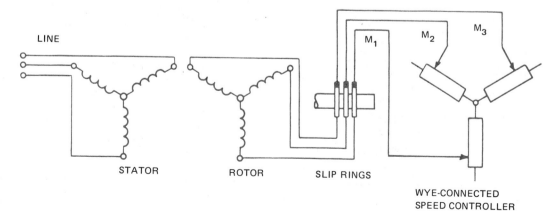

Fig. 17–5 Connections for a wound-rotor induction motor and a speed controller

Speed Control

The insertion of resistance in the rotor circuit not only limits the starting surge of current, but also produces a high starting torque and provides a means of adjusting the speed. If the full resistance of the speed controller is inserted into the rotor circuit when the motor is running, the rotor current decreases and the motor slows down. As the rotor speed decreases, more voltage is induced in the rotor windings and more rotor current is developed to create the necessary torque at the reduced speed.

If all of the resistance is removed from the rotor circuit, the current and the motor speed will increase. However, the rotor speed always will be less than the synchronous speed of the field developed by the stator windings. Recall that this fact also is true of the squirrel-cage induction motor. The speed of a wound-rotor motor can be controlled manually or automatically with timing relays, contactors, and pushbutton speed selection.

Torque Performance

As a load is applied to the motor, both the percent slip of the rotor and the torque developed in the rotor increase. As shown in the graph in figure 17–6, the relationship between the torque and percent slip is practically a straight line.

Figure 17–6 illustrates that the torque performance of a wound-rotor induction motor is good whenever the full resistance of the speed controller is inserted in the rotor

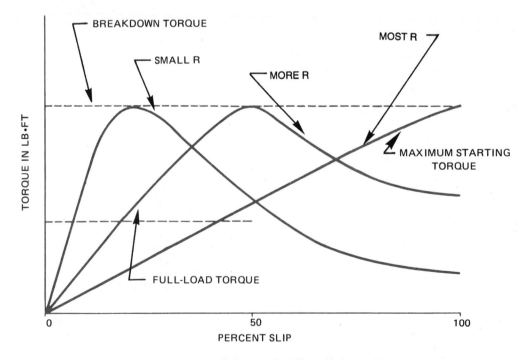

Fig. 17–6   Performance curves of a wound-rotor motor

circuit. The large amount of resistance in the rotor circuit causes the rotor current to be almost in phase with the induced voltage of the rotor. As a result, the field set up by the rotor current is almost in phase with the stator field. If the two fields reach a maximum value at the same instant, there will be a strong magnetic reaction resulting in a high torque output.

However, if all of the speed controller resistance is removed from the rotor circuit and the motor is started, the torque performance is poor. The rotor circuit minus the speed controller resistance consists largely of inductive reactance. This means that the rotor current lags behind the induced voltage of the rotor and, thus, the rotor current lags behind the stator current. As a result, the rotor field set up by the rotor current lags behind the stator field which is set up by the stator current. The resulting magnetic reaction of the two fields is relatively small since they reach their maximum values at different points. In summary, then, the starting torque output of a wound-rotor induction motor is poor when all resistance is removed from the rotor circuit.

Speed Regulation

It was shown in the previous paragraphs that the insertion of resistance at the speed controller improves the starting torque of a wound-rotor motor at low speeds. However, there is an opposite effect at normal speeds. In other words, the speed regulation of the motor is poorer when resistance is added in the rotor circuit at a higher speed. For this reason, the resistance of the speed controller is removed as the motor comes up to its rated speed.

Figure 17–7 shows the speed performance of a wound-rotor induction motor. Note that the speed characteristic curve resulting when all of the resistance is cut out of the speed controller indicates relatively good speed regulation. The second speed characteristic curve, resulting when all of the resistance is inserted in the speed controller, has a marked drop in speed as the load increases. This indicates poor speed regulation.

Power Factor

The power factor of a wound-rotor induction motor at no load is as low as 15 percent to 20 percent lag. However, as load is applied to the motor, the power factor improves and increases to 85 percent to 90 percent lag at rated load.

Figure 17–8 is a graph of the power factor performance of a wound-rotor induction motor from a no-load condition to full load. The low lagging power factor at no load is due to the fact that the magnetizing component of load current is such a large part of the total motor current. The magnetizing component of load current magnetizes the iron, causing interaction between the rotor and the stator, by mutual inductance.

As the mechanical load on the motor increases, the in-phase component of current increases to supply the increased power demands. The magnetizing component of the cur-

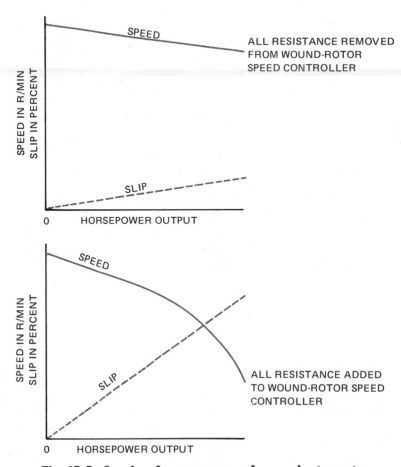

Fig. 17–7 Speed performance curves of a wound-rotor motor

rent remains the same, however. Since the total motor current is now more nearly in phase with the line voltage, there is an improvement in the power factor.

Operating Efficiency

Both a wound-rotor induction motor with all of the resistance cut out of the speed controller and a squirrel-cage induction motor show nearly the same efficiency performance. However, when a motor must operate at slow speeds with all of the resistance cut in the rotor circuit, the efficiency of the motor is poor because of the power loss in watts in the resistors of the speed controller.

Figure 17–9 illustrates the efficiency performance of a wound-rotor induction motor. The upper curve showing the highest operating efficiency results when the speed controller is in the fast position and there is no resistance inserted in the rotor circuit. The lower curve shows a lower operating efficiency. This occurs when the speed controller is in the slow position and all of the controller resistance is inserted in the rotor circuit.

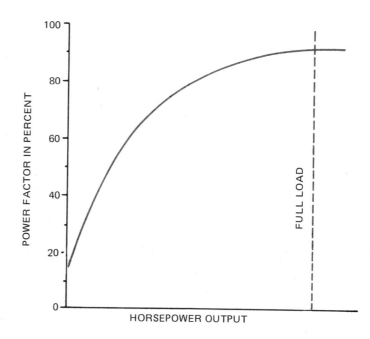

Fig. 17–8   Power factor of a wound-rotor induction motor

Reversing Rotation

The direction of rotation of a wound-rotor induction motor is reversed by interchanging the connections of any two of the three line leads, figure 17–10. This procedure is identical to the procedure used to reverse the direction of rotation of a squirrel-cage induction motor.

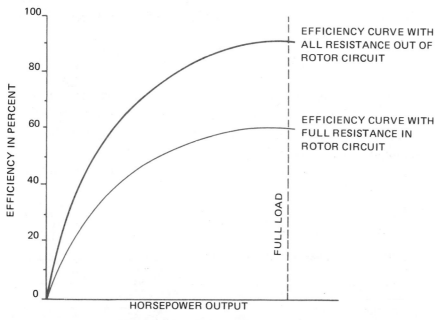

Fig. 17–9   Efficiency curves for a wound-rotor induction motor

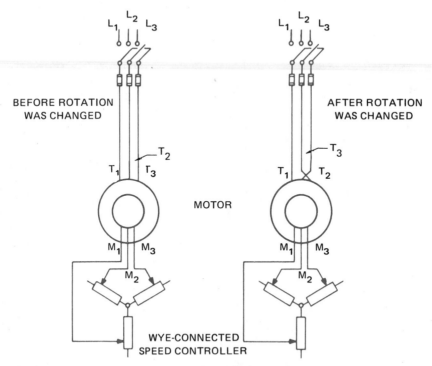

Fig. 17–10   Changes necessary to reverse direction of rotation of a wound-rotor motor

The electrician should never attempt to reverse the direction of rotation of a wound-rotor induction motor by interchanging any of the leads feeding from the slip rings to the speed controller. Changes in these connections will not reverse the direction of rotation of the motor.

SUMMARY

The wound-rotor motor is rarely installed as a new motor today, but there are still a number of the motors in use. The wound rotor motor may be used for variable speed with the insertion of secondary resistors. The starting current and starting torque of the motor were the prime considerations when selecting the wound-rotor motor for installation. There are still many references to the wound-rotor motor used in the National Electrical Code.

ACHIEVEMENT REVIEW

Give complete answers to the following questions.

1.   List the essential parts of a wound-rotor induction motor. _____

_____

_____

2. List two reasons why a wound-rotor induction motor is started with all of the resistance inserted in the speed controller. _____

_____

_____

_____

3. A three-phase, wound-rotor induction motor has six poles and is rated at 60 hertz. The full-load speed of this motor with all of the resistance cut out of the speed controller is 1,120 r/min. What is the synchronous speed of the field set up by the stator windings? _____

_____

_____

_____

4. Determine the percent slip at the rated load for the motor in question 3.

_____

_____

_____

5. Why is a wound-rotor induction motor used in place of a squirrel-cage induction motor for some industrial applications?_____

_____

_____

_____

6. Why is the percent efficiency of a wound-rotor induction motor poor when operating at rated load with all of the resistance inserted in the speed controller?

_____

_____

_____

7. What must be done to reverse the direction of rotation of a wound-rotor induction motor? _____

_____

_____

8. Why is the power factor of a wound-rotor induction motor poor at no load?

_____

_____

9. List the two factors which affect the synchronous speed of the rotating magnetic field set up by the current in the stator windings. _____

_____

_____

B. Select the correct answer for each of the following statements and place the corresponding letter in the space provided.

10. The speed of a wound-rotor motor is increased by   _____
    a. inserting resistance in the primary circuit.
    b. inserting resistance in the secondary circuit.
    c. decreasing the resistance in the secondary circuit.
    d. decreasing the resistance in the primary circuit.

11. The starting current of a wound-rotor induction motor is limited by   _____
    a. decreasing the resistance in the primary circuit.
    b. decreasing resistance in the secondary circuit.
    c. inserting resistance in the primary circuit.
    d. inserting resistance in the secondary circuit.

12. The direction of rotation of a wound-rotor motor is changed by
    interchanging any two of the three: _____
    a. L_1, L_2 , or L_3.   c. M_1, M_2, or M_3.
    b. T_1, T_2, or T_3.   d. all of these.

13. Wound-rotor motors can be used with   _____
    a. manual speed controllers.
    b. automatic speed controllers.
    c. pushbutton selection.
    d. all of these.

14. The full-load efficiency of a wound-rotor motor is best when   _____
    a. all of the resistance is cut out of the secondary circuit.
    b. all of the resistance is cut in the secondary circuit.
    c. it is running slowly.
    d. it is running at medium speed.

15. The main advantage of the wound-rotor polyphase motor is that it   _____
    a. has a low starting torque.   c. will reverse rapidly.
    b. has a wide speed range.   d. has a low speed range.

16. The wound-rotor motor is so-named because the   _____
    a. rotor is wound with wire.
    b. stator is wound with wire.
    c. controller is wound with wire.
    d. all of these.

17. The magnetizing component of load current   _____
    a. is a small part of the total motor current at no load.
    b. magnetizes the iron, causing interaction between the rotor and
       the stator.
    c. is a large part of the total motor current at full load.
    d. is unrelated to the power factor.

18

MANUAL SPEED CONTROLLERS FOR WOUND-ROTOR INDUCTION MOTORS

OBJECTIVES

After studying this unit, the student will be able to

- state three reasons why speed controllers are used with wound-rotor induction motors.

- list and describe the physical construction of two basic types of manual speed controllers.

- explain the operation of a faceplate controller with a protective starting device.

- explain the operation of a drum controller.

- summarize the National Electrical Code regulations regarding the wire size for the stator circuit, the wire size for the rotor circuit, starting overload protection, and running overload protection.

- draw wiring diagrams for wound rotor motor control.

Manual speed controls for wound rotor motors are quickly becoming obsolete. However, the maintenance electrician still needs a working knowledge of the operations of this type of control.

Many industrial applications require the use of wound-rotor induction motors with speed controllers. This unit covers the details of the operation of manual speed controllers and their connection to wound-rotor motors. Information on National Electrical Code regulations which apply to wound-rotor induction motor installations also is presented.

Reasons for Use of Speed Controllers

Speed controllers are used with wound-rotor induction motors for three basic reasons:

- to limit the starting surge of current to the motor by inserting resistance in the rotor circuit.

- to improve the starting torque of a wound-rotor induction motor by inserting resistance in the rotor circuit.
- to control the speed of a wound-rotor induction motor by varying the resistance in the rotor circuit.

FACEPLATE CONTROLLER

The simplest form of manual speed controller is the *faceplate controller* (figure 18–1). In this type of controller, three sets of contact buttons are mounted on a panel. Each set of contact buttons is connected to a separate tapped resistor housed in the speed controller box. The resistance value of each resistor section is varied by a contact arm. For the faceplate controller shown in figure 18–l, note that the three contact arms are connected at a common point at the center. A handle attached to one of the arms moves all arms simultaneously. These arms are spaced 120 mechanical degrees apart so that equal amounts of resistance can be cut in or out of each tapped resistor. This faceplate manual speed controller is connected in a three-phase wye arrangement.

Figure 18–2 shows the connections between a faceplate speed controller and a wound-rotor induction motor. As the three-phase, wound-rotor motor is started, all of the resistance in the speed controller is inserted in the rotor circuit. The stator circuit is connected across the three-phase line voltage by an across-the-line motor starter switch controlled from a pushbutton station. Because the maximum value of resistance is inserted in the rotor circuit at startup, the starting surge of current is limited. As a result, the starting torque is improved. After the motor has started, resistance is cut out of the rotor circuit using the speed controller until the desired speed is obtained.

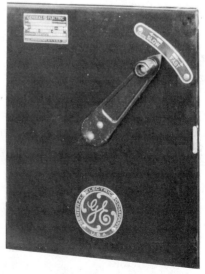

Fig. 18–1   Speed regulating rheostat with auxiliary control switch for interlocking with or controlling operation of a magnetic line switch (*Photo courtesy of General Electric Company*)

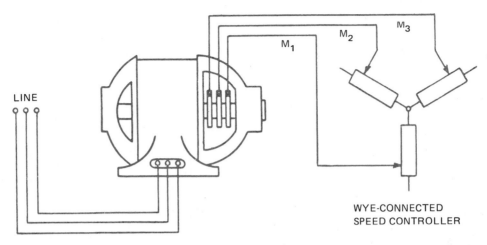

Fig. 18–2   Connections of a faceplate speed controller to a wound-rotor motor

Faceplate Controller with Protective Starting Device

If a wound-rotor induction motor is started with all of the speed controller resistance cut out of the rotor circuit, the starting surge of current to the stator windings will be high and the starting torque developed by the motor will be poor.

A protective starting device is used with a faceplate controller to insure that the motor is started with all of the resistance inserted (figure 18–3). The motor can be started only when the arms of this special type of controller are in the slow position with all of the resistance inserted in the rotor circuit.

A speed controller of this type has a pair of contacts which are closed when the three movable arms of the controller are in the slow position. These contacts are in series with the normally open start pushbutton. When the start pushbutton is pressed, the coil of the magnetic across-the-line starter switch is energized. The starter contacts close and the rated three-phase line voltage is applied to the stator windings. As the motor accelerates, the movable arms of the speed controller can be adjusted to obtain the desired speed. As the arms are moved from the slow position, the faceplate controller circuit contacts open. However, the contacts of the across-the-line magnetic motor switch are closed. Since one pair of these contacts acts as a sealing circuit around the normally open pushbutton and the open contacts of the circuit on the faceplate controller, the main starter coil remains energized and the motor continues to operate.

The motor cannot be started if the speed controller is not in the slow position. This is due to the fact that the control circuit to the coil of the magnetic motor switch is open because the contacts of the faceplate controller are open. Therefore, the motor switch will not operate when the start pushbutton is closed. To start the motor, the adjustable arms of the faceplate controller must be in the start position.

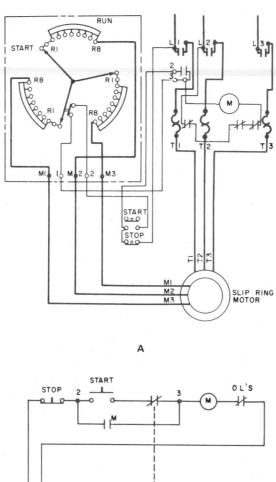

A

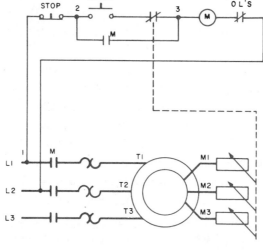

B

Fig. 18–3   Faceplate controller with a protective starting device A) Wiring diagram of a manual speed regulator interlocked with a magnetic starter B) Elementary diagram of figure 18–3A (From Alerich, *Electric Motor Control,* **4th Edition, copyright 1988 by Delmar Publishers.)**

DRUM CONTROLLER

The drum controller, figure 18–4, is another type of manual speed controller which can be used with wound-rotor induction motors.

A drum controller consists basically of a case, contact fingers, the cylinder assembly, and external resistors. The case consists of a back piece and end pieces made of plate metal and a cover made of sheet metal. The cover fits over the end pieces and is removed when maintenance or repair work is required. The cover may be provided with a rubber gasket to make it dustproof. The wiring is brought to the controller through bushed holes or condulet fittings either in the back plate or the bottom end piece of the case.

The contact fingers of the drum controller are stationary contacts. The three wires from the rotor slip rings are connected to these contacts, as are the wires from the externally mounted grid resistors. Each contact finger is made of brass or steel and has a copper tip. Each finger is mounted and pivoted so that adjustments can be made on a spring to obtain the proper contact tension.

A drum controller also contains a vertical cylinder mounted on an insulated shaft. This cylinder provides the moving contacts that make and break connections to the various fixed contact fingers. The contacts on the cylinder are made of rolled copper segments which are moved by a handle located at the top of the speed controller.

Fig. 18–4 Drum controller used for control of a wound rotor motor

The controller resistors usually are mounted outside and behind the controller case. The resistor units are cast from iron or a metal alloy, figure 18–5. Connecting wires from taps on the resistors terminate at contact fingers in the drum controller.

Operation of a Drum Controller

When the controller handle is in the slow position, the maximum resistance of the controller is inserted in the three phases of the rotor circuit. (The resistors of the speed controller are connected in wye.) As the controller handle is moved toward the run position, the copper contacts on the cylinder assembly make contact with various stationary fingers and sections of the resistance are cut out of the rotor circuit. When the handle is in

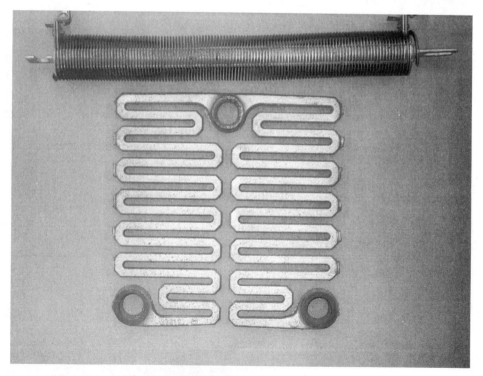

Fig. 18–5   Resistors that may be used in wound rotor motor control (From Keljik, *Electric Motors and Motor Controls,* **copyright 1995 by Delmar Publishers)**

the run position, all of the resistance of the speed controller is cut out of the rotor circuit. As a result, the motor operates at its rated speed.

Figure 18–6 is a diagram of the internal connections of a drum-type manual speed controller. This circuit indicates that in the run position, the leads from the rotor slip rings are connected together and the resistance of the controller is cut out of the rotor circuit.

The stator circuit is connected directly across the three-phase line by an across-the-line motor switch controlled from a pushbutton station. Some drum controllers may have a small pair of main relay coil contacts which are closed only when the speed controller is in the start position. These contacts are in series with the normally open start pushbutton. If an attempt is made to start the motor with the speed controller handle not in the slow position, the motor will not start. If the speed controller handle is returned to the slow position and the start pushbutton pressed, the motor will accelerate slowly. The operator can then adjust the speed of the motor to the desired value. After the motor has started, the coil contacts and the normally open start pushbutton is shunted out by the sealing contactors in the magnetic motor switch.

NATIONAL ELECTRICAL CODE REGULATIONS

The National Electrical Code in *Section 430-23* requires for continuous duty that the conductors from the rotor slip rings to the speed controller have an ampacity (current-carrying capacity) not less than 125 percent of the full-load current rating of the rotor, secondary circuit.

The Code lists several special industrial applications where the resistors are separate from the controller for which percentage values of the full load current other than 125 percent are allowed for the determination of wire size (the conductors connecting the controller to the resistors). As in the previous paragraph, for motors used in special industrial applications, the Code permits the application of other percentage values to the full-load-current rating of the rotor to determine the rotor circuit wire size.

If the speed controller resistors are separated from the speed controller, conductors are required to feed from the connection points on the resistors to the contact fingers. The

Resistor Duty Classification	Ampacity of Wire in Percent of Full-load Rotor Current
Light starting	35
Heavy starting	45
Extra heavy starting	55
Light intermittent	65
Medium intermittent	75
Heavy intermittent	85
Continuous	110

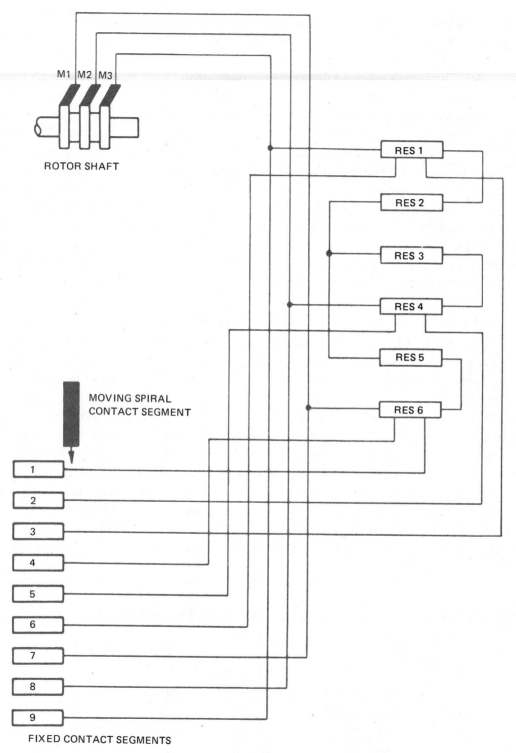

Fig. 18-6 Connections of a drum-type manual controller

Code also gives specific information on the percentage values of full-load current to be used to determine the size of these conductors.

The fusible starting protection for a wound-rotor induction motor is to be not more than 150 percent of the full-load motor current rating (NEC *Table 430-152*). This starting protection usually consists of fuses located in the motor disconnect switch. Wire ampacity is no less than 125 percent of the motor full load current (NEC *Section 430-22*).

Running overload protection is to be provided for wound-rotor induction motors. For a motor rated at more than one hp and marked to have a temperature rise of not more than 40 degrees Celsius, the running overload protection shall be rated at not more than 125 percent of the motor full-load current rating [NEC *Section 430-32(a)*]. The running overload protection usually consists of thermal overload units located in the magnetic motor starter.

The Code also states that the secondary circuits of wound-rotor induction motors, including conductors, controller, and resistors, shall be considered to be protected by the running overload devices provided in the primary circuit. [See NEC *Section 430-32(d)*.]

The phrase *primary circuit,* when referring to wound-rotor induction motors, means the stator winding. The phrase *secondary circuit,* as applied to wound-rotor induction motors, means the rotor circuit.

TERMINAL MARKINGS

The stator leads of a, three-phase, wound-rotor induction motor are marked T_1 T_2, and T_3. Note that this is the marking system used with three-phase, squirrel-cage induction motors.

The rotor leads of a wound-rotor motor are marked M_1, M_2, and M_3. The M_1 lead connects to the slip ring nearest the bearing housing, lead M_2 connects to the middle slip ring, and lead M_3 connects to the slip ring nearest the rotor windings.

SUMMARY

Manual speed controllers are still in operation because of their ruggedness and reliability. The concept of manually controlling the wound-rotor motor is to control the primary (stator) circuit and to control the resistance of the secondary (rotor) circuit. It is important to be able to understand basic control philosophy and to read diverse control schemes. A working knowledge of the wound-rotor motor principle helps in all types of motor theory and control system troubleshooting.

ACHIEVEMENT REVIEW

A.   Give complete answers to the following questions.

1.   Give three reasons why speed controllers are used with wound-rotor induction motors. _____

_____

_____

2. List the two basic types of manual speed controllers used with wound-rotor induction motors. _____

_____

3. A 230-volt, three-phase, 10-horsepower wound-rotor induction motor has a fullload stator current rating of 28 amperes and a full-load rotor current rating of 40 amperes per terminal. Determine the fuse size required for starting overload protection for this induction motor. _____

_____

_____

4. Determine the size of the thermal overload units required for the induction motor given in question 2. _____

_____

_____

5. What size wire, Type THHN, should be used for the stator circuit of the motor given in question 2?_____

_____

_____

6. What size wire, Type THHN, should be used for the connections between the slip rings and the speed controller for the motor in question 3? _____

_____

_____

7. What size wire, Type THHN, should be used between the external resistors and the speed controller in the motor in question 3 if the duty classification is "Light Intermittent Duty"? Check the Code before answering this problem.

_____

_____

_____

_____

8. Draw the wiring connection diagram of a wound-rotor induction motor which is started by means of an across-the-line magnetic motor starter switch controlled from a pushbutton station. Include a wye-connected speed controller in the rotor circuit.

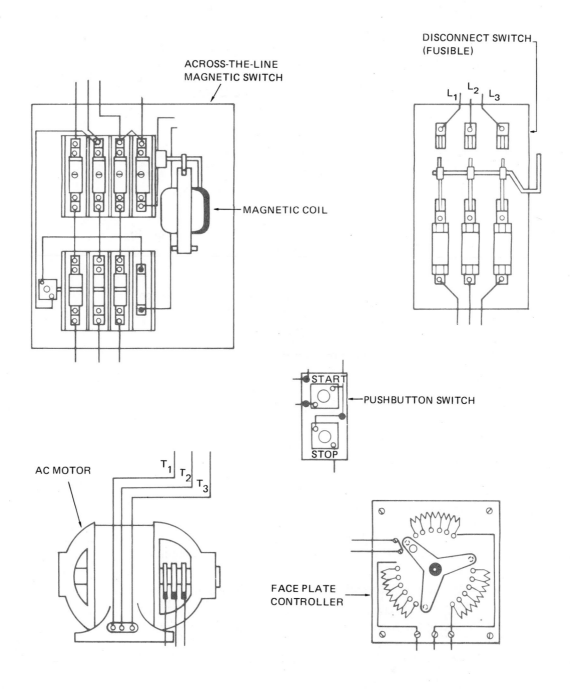

9. What precaution is provided on some speed controllers to prevent the starting of the motor when the speed controller is left in the run position? _____

_____

_____

_____

B. Select the correct answer for each of the following statements and place the corresponding letter in the space provided.

10. What type of speed controller is used with a wound-rotor motor? _____
   a. Primary resistor   c. Autotransformer
   b. Secondary resistor   d. Reactor

11. When starting a wound-rotor motor, full resistance is inserted in the secondary circuit and a magnetic starter applies full line voltage to _____
   a. the stator windings.   c. the starting resistors.
   b. the slip wings.   d. the primary resistors.

12. The rotor winding leads of a wound-rotor motor are brought out to _____
   a. two slip rings.   c. four slip rings.
   b. three slip rings.   d. a centrifugal switch.

13. The stator circuit of a wound-rotor motor is sometimes called the _____
   a. primary circuit.   c. rotor.
   b. secondary circuit.   d. frame.

14. The terminal lead connected to the slip ring nearest the bearing housing is _____
   a. T_1.   c. M_3.
   b. T_3.   d. M_1.

15. A drum controller is used for _____
   a. reversing the stator.
   b. switching resistances in the secondary.
   c. braking.
   d. reversing the rotor.

16. For best starting results, apply _____
   a. minimum voltage to the stator, with maximum resistance in the rotor.
   b. full voltage to the stator, with maximum resistance in the rotor.
   c. full voltage to the rotor, with maximum resistance in the stator.
   d. full voltage to the stator, with minimum resistance in the rotor.

U • N • I • T
19

SUMMARY REVIEW OF UNITS 13-18

OBJECTIVE

- To give the student an opportunity to evaluate the knowledge and understanding acquired in the study of the previous six units.

1. List three types of three-phase ac motors.

    a. _____

    b. _____

    c. _____

2. Insert the word or phrase to complete each of the following statements.

    a. The speed of a three-phase induction motor falls slightly from a no-load condition to a full load. This is true of a three-phase induction motor with a _____ _____ rotor.

    b. A speed controller is used only with a three-phase induction motor of the _____ type.

    c. When all of the resistance of the speed controller is inserted in the secondary circuit of a three-phase _____ _____ induction motor, the starting torque is very good.

    d. A three-phase _____ motor is operated with an overexcited dc field to obtain a leading power factor.

    e. The speed of a three-phase _____ motor remains constant from a no-load condition to full load if the operating frequency remains constant.

3. State two advantages of using a squirrel-cage induction motor. _____

    _____

    _____

    _____

4. State one disadvantage of using a squirrel-cage induction motor._____

    _____

    _____

5. Explain how the direction of rotation of a three-phase, squirrel-cage induction motor is reversed. _____

_____

_____

_____

6. A two-pole, 60-Hz, three-phase, squirrel-cage induction motor has a full-load speed of 3,475 r/min. Determine the synchronous speed of this motor. _____

_____

_____

7. Determine the percent slip of the motor in question 6. _____

_____

_____

8. What is the purpose of starting protection for a three-phase motor?

_____

_____

_____

9. What is the purpose of running protection for a three-phase motor? _____

_____

_____

10. Show the connection diagram for the nine terminal leads of a wye-connected, three-phase motor rated at 230/460 volts for three-phase operation on 230 volts.

11. Explain how the running over load protection for a three-phase motor rated at more than 1 horsepower would be selected. _____

_____

_____

_____

12. Insert the correct code letter in the following statements.
    a. The National Electrical Code requires squirrel-cage induction motors with code marking _____ to have starting protection rated at not over 150 percent of full-load current for a nontime-delay fuse.
    b. The National Electrical Code requires squirrel-cage induction motors with code letter markings _____ to have starting protection rated at not over 250 percent of full-load current for a nontime-delay fuse.
    c. The National Electrical Code requires squirrel-cage induction motors with _____ code letter markings to have starting protection rated at not over 300 percent of full-load current for a nontime-delay fuse.
    d. The National Electrical Code requires squirrel-cage induction motors with autotransformer starting and code letter markings _____ to have starting protection rated at not over 200 percent of full-load current for a nontime-delay fuse.

13. Why are starting compensators used with large three-phase, squirrel-cage induction motors?_____

_____

_____

14. Why is a wound-rotor induction motor used in place of a squirrel-cage induction motor for some industrial applications?_____

_____

_____

_____

_____

15. Explain how the direction of rotation of a three-phase, wound-rotor induction motor can be reversed. _____

_____

_____

_____

16. Insert the correct word or phrase to complete each of the following statements.
    a. The speed of a wound-rotor induction motor is _____ by inserting resistance in the rotor circuit through a speed controller.
    b. The starting surge of current of a wound-rotor induction motor is minimized by

    _____ .

    c. The rotation of a wound-rotor induction motor is _____ by changing any two of the three leads feeding from the rotor slip rings to the speed controller.
    d. The _____ of a wound-rotor induction motor is very good if all of the resistance of the speed controller is inserted in the rotor circuit.
    e. The efficiency of a wound-rotor induction motor operating at rated load with all of the resistance inserted in the rotor circuit is _____ .

17. Draw a schematic connection diagram of a wound-rotor induction motor which is started by means of an across-the-line magnetic motor starter controlled from a pushbutton station. Include a wye-connected speed controller in the rotor circuit.

18. Explain how the direction of rotation of a three-phase synchronous motor is reversed. _____

_____

_____

19. List two important applications for three-phase synchronous motors.

_____

_____

20. A three-phase synchronous motor with four stator poles and four rotor poles is oper-ated from a three-phase, 60-Hz line of the correct voltage rating. Determine the speed of the motor._____

_____

_____

_____

21. Explain the correct procedure for starting a three-phase synchronous motor.

_____

_____

_____

22. Insert the correct word or phrase to complete each of the following statements.
    a. The speed of a synchronous motor is _____ from no load to full load.
    b. A synchronous motor with an underexcited dc field has a _____ power factor.
    c. A three-phase _____ motor must be started as an induction motor.

23. What is an amortisseur winding? _____

_____

_____

_____

24. Explain what is meant by the term jogging._____

_____

_____

_____

25. Explain what is meant by the term plugging._____

_____

_____

26. What identifying information should appear on a motor controller to comply with the requirements of the National Electrical Code? _____

_____

_____

_____

_____

27. A three-phase, squirrel-cage induction motor is rated at 25 hp, 230 volts, 64 amperes per terminal, 40 degrees Celsius, and is classified as code letter F. In the following table, fill in the correct fuse size required for branch-circuit protection and the correct running overload protection for this motor, when used with each of the types of controllers listed.

| | Fuse Protection | | Running Overcurrent Protection |
Type of Controller	Nontime-delay	Time-delay	
a. Resistance starter			
b. Automatic autotransformer compensator			
c. Across-the-line magnetic motor starting switch with jogging capability			
d. Across-the-line magnetic motor starting switch with plugging capability			

28. What size of wire and conduit are used for the branch circuit feeding the motor in question 28? (Use Type TW.) _____

_____

_____

_____

29. Insert the correct word or phrase to complete each of the following statements.
    a. A controller with _____ may be used to stop a motor quickly.
    b. When fuses are used as protection for a three-phase, three-wire ungrounded branch motor circuit, the fuses must be installed in _____ line leads.
    c. Motors operating on a three-phase, three-wire ungrounded system require _____ thermal overload units for running overcurrent protection.

30. How is dynamic braking applied to an induction motor? _____

_____

_____

_____

_____

31. How is dynamic braking applied to a synchronous motor?_____

_____

_____

_____

_____

32. What is a megohmmeter?_____

_____

_____

_____

_____

33. Draw a schematic diagram of an across-the-line magnetic switch connected to a three-phase, squirrel-cage induction motor. The magnetic switch has jogging capability. Include in the connection diagram the main relay coil, the pushbutton station with start, jog, and stop pushbuttons, and the maintaining contact.

34. Draw a schematic diagram of a wye-delta starter, complete with pushbutton station, connected to a three-phase motor.

35. On periodic tests, a motor winding suddenly drops to a low resistance value. Testing with a _____ determines this condition.

36. How is a megohmmeter used to measure the insulation resistance of the windings of an ac motor?_____

   _____

   _____

   _____

   _____

   _____

   _____

37. What is a growler?_____

   _____

   _____

   _____

   _____

   _____

   _____

38. Explain how a growler is used to locate a short-circuit condition in a motor winding.

_____

_____

_____

_____

_____

_____

_____

_____

_____

39. For the sleeve bearings of an ac motor, explain how the old oil is removed and the
bearings cleaned and lubricated. _____

_____

_____

_____

_____

_____

_____

_____

_____

_____

40. Place the correct answers in each of the spaces provided in the following diagram. Refer to the National Electrical Code.

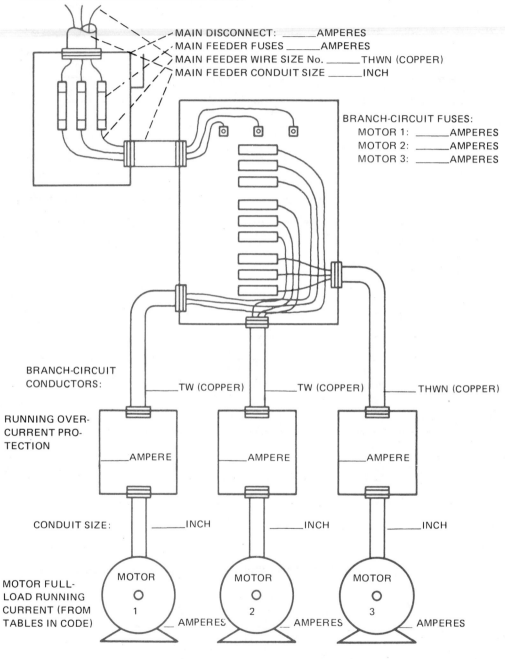

MAIN DISCONNECT: _____ AMPERES
MAIN FEEDER FUSES _____ AMPERES
MAIN FEEDER WIRE SIZE No. _____ THWN (COPPER)
MAIN FEEDER CONDUIT SIZE _____ INCH

BRANCH-CIRCUIT FUSES:
  MOTOR 1: _____ AMPERES
  MOTOR 2: _____ AMPERES
  MOTOR 3: _____ AMPERES

BRANCH-CIRCUIT
CONDUCTORS: _____ TW (COPPER)   _____ TW (COPPER)   _____ THWN (COPPER)

RUNNING OVER-
CURRENT PRO-
TECTION
              _____ AMPERE   _____ AMPERE   _____ AMPERE

CONDUIT SIZE: _____ INCH   _____ INCH   _____ INCH

MOTOR FULL-
LOAD RUNNING   MOTOR 1   MOTOR 2   MOTOR 3
CURRENT (FROM
TABLES IN CODE) _____ AMPERES   _____ AMPERES   _____ AMPERES

5 HP 230 VOLT,   15 HP 230 VOLT,   30 HP 230 VOLT,
THREE-PHASE   THREE-PHASE   THREE-PHASE
FULL VOLTAGE START   FULL VOLTAGE START   AUTOTRANSFORMER START
CODE LETTER K   CODE LETTER J   CODE LETTER B

THE SYNCHRONOUS MOTOR

OBJECTIVES

After studying this unit, the student will be able to

- list the basic parts in the construction of a synchronous motor.

- define and describe an amortisseur winding.

- describe the basic operation of a synchronous motor.

- describe how the power factor of a synchronous motor is affected by an underexcited dc field, a normally excited dc field, and an overexcited dc field.

- list at least three industrial applications of the synchronous motor.

The *synchronous motor*, figure 20–1, is a three-phase ac motor which operates at a constant speed from a no-load condition to full load. This type of motor has a revolving field which is separately excited from a direct-current source. In this respect, it is similar

Fig. 20–1 Synchronous motor with direct-connected exciter (*Photo courtesy of General Electric Company*)

to a three-phase ac generator. If the de field excitation is changed, the power factor of a synchronous motor can be varied over a wide range of lagging and leading values.

The synchronous motor is used in many industrial applications because of its fixed speed characteristic over the range from no load to full load. This type of motor also is used to correct or improve the power factor of three-phase ac industrial circuits, thereby reducing operating costs.

CONSTRUCTION DETAILS

A three-phase synchronous motor basically consists of a stator core with a three-phase winding (similar to an induction motor) a revolving dc field with an auxiliary or amortisseur winding and slip rings, brushes and brush holders, and two end shields housing the bearings that support the rotor shaft. An *amortisseur winding* (figure 20–2) consists of copper bars embedded in the cores of the poles. The copper bars of this special type of "squirrel-cage winding" are welded to end rings on each side of the rotor.

Both the stator winding and the core of a synchronous motor are similar to those of the three-phase, squirrel-cage induction motor and the wound-rotor induction motor. The leads for the stator winding are marked T_1, T_2, and T_3 and terminate in an outlet box mounted on the side of the motor frame.

The rotor of the synchronous motor has salient field poles. The field coils are connected in series for alternate polarity. The number of rotor field poles must equal the number of stator field poles. The field circuit leads are brought out to two slip rings mounted on the rotor shaft for brush-type motors. Carbon brushes mounted in brush holders make contact with the two slip rings. The terminals of the field circuit are brought out from the

AMORTISSEUR
(SQUIRREL-CAGE
WINDING)

SLIP RINGS

SALIENT POLES

Fig. 20–2   A synchronous motor rotor with amortisseur winding (*Photo courtesy of General Electric Company*)

brush holders to a second terminal box mounted on the frame of the motor. The leads for the field circuit are marked F_1 and F_2. A squirrel-cage, or amortisseur, winding is provided for starting because the synchronous motor is not selfstarting without this feature. The rotor shown in figure 20–2 has salient poles and an amortisseur winding.

Two end shields are provided on a synchronous motor. One of the end shields is larger than the second shield because it houses the dc brush holder assembly and slip rings. Either sleeve bearings or ball-bearing units are used to support the rotor shaft. The bearings also are housed in the end shields of the motor.

PRINCIPLE OF OPERATION

When the rated three-phase voltage is applied to the stator windings, a rotating magnetic field is developed. This field travels at the synchronous speed. As stated in previous units, the synchronous speed of the magnetic field depends on the frequency of the three-phase voltage and the number of stator poles. The following formula is used to determine the synchronous speed.

$$\text{Synchronous speed} = \frac{120 \times \text{frequency}}{\text{number of poles}}$$

$$S = \frac{120 \times f}{p}$$

The magnetic field which is developed by the stator windings travels at synchronous speed and cuts across the squirrel-cage winding of the rotor. Both voltage and current are induced in the bars of the rotor winding. The resulting magnetic field of the amortisseur (squirrel-cage) winding reacts with the stator field to create a torque which causes the rotor to turn.

The rotation of the rotor will increase in speed to a point slightly below the synchronous speed of the stator, about 92 percent to 97 percent of the motor rated speed. There is a small slip in the speed of the rotor behind the speed of the magnetic field set up by the stator. In other words, the motor is started as a squirrel-cage induction motor.

The field circuit is now connected to a source of direct current and fixed magnetic poles are set up in the rotor field cores. The magnetic poles of the rotor are attracted to unlike magnetic poles of the magnetic field set up by the stator.

Figures 20-3 and 20-4 show how the rotor field poles lock with unlike poles of the stator field. Once the field poles are locked, the rotor speed becomes the same as the speed of the magnetic field set up by the stator windings. In other words, the speed of the rotor is now equal to the synchronous speed.

Remember that a synchronous motor must always be started as a three-phase, squirrel-cage induction motor with the dc field excitation disconnected. The dc field circuit is added only after the rotor accelerates to a value near the synchronous speed. The motor then will operate as a synchronous motor, locked in step with the stator rotating field.

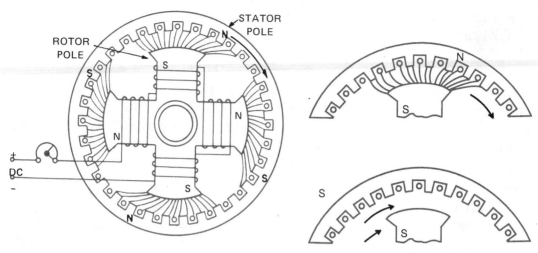

Fig. 20–3   Diagram to show the principle of operation on a synchronous motor

REPULSION BETWEEN UNLIKE POLES

Fig. 20–4   Starting of synchronous motors

If an attempt is made to start a three-phase synchronous motor by first energizing the dc field circuit and then applying the three-phase voltage to the stator windings, the motor will not start since the net torque is zero, At the instant the three-phase voltage is applied to the stator windings, the magnetic field set up by the stator current turns at the synchronous speed. The rotor, with its magnetic poles of fixed polarity, is attracted first by an unlike stator pole and attempts to turn in that direction. However, before the rotor can turn, another stator pole of opposite polarity moves into position and the rotor then attempts to turn in the opposite direction. Because of this action of the alternating poles, the net torque is zero and the motor does not start.

Direct-Current Field Excitation

In the early models, the field circuit is excited from an external direct-current source. A dc generator may be coupled to the motor shaft to supply the dc excitation current.

Figure 20–5 shows the connections for a synchronous motor. A field rheostat in the separately excited field circuit varies the current in the field circuit. Changes in the field current affect the strength of the magnetic field developed by the revolving rotor. Variations in the rotor field strength do not affect the motor which continues to operate at a constant speed. However, changes in the dc field excitation do change the power factor of a synchronous motor.

Brushless Solid-State Excitation

An improvement in synchronous motor excitation is the development of the brushless dc exciter. The commutator of a conventional direct-connected exciter is replaced

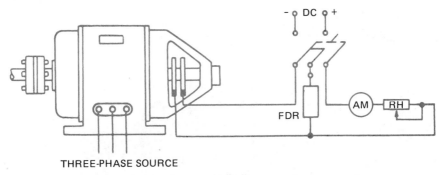

Fig. 20–5   External connections for a synchronous motor

with a three-phase, bridge-type, solid-state rectifier. The dc output is then fed directly to the motor field winding. Simplified circuitry is shown in figure 20–6. A stationary field ring for the ac exciter receives dc from a small rectifier in the motor control cabinet. This rectifier is powered from the ac source. The exciter dc field is also adjustable. Rectifier solid-state diodes change the exciter ac output to dc. This dc is the source of excitation for the rotor field poles. Silicon-controlled rectifiers, activated by the solid-state field control circuit, replace electromechanical relays and the contactors of the conventional brush-type synchronous motor.

The field discharge resistor is inserted during motor starting. At motor synchronizing pull-in speed, the field discharge circuit is automatically opened and dc excitation is applied to the rotor field pole windings. Excitation is automatically removed if the motor pulls out of step (synchronization) due to an overload or a voltage failure. The brushless rotor is shown in figure 20–7. Mounted on the rotor shaft are the armature of the ac exciter, the ac output of which is rectified to dc by the silicon diodes. Brush and commutator problems are eliminated with this system. (The stator of a brushless motor is similar to that of a brush-type motor.)

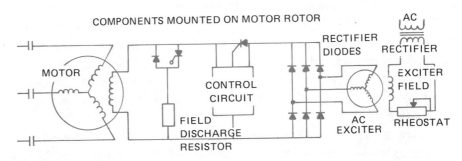

Fig. 20–6   Simplified circuit for a brushless synchronous motor

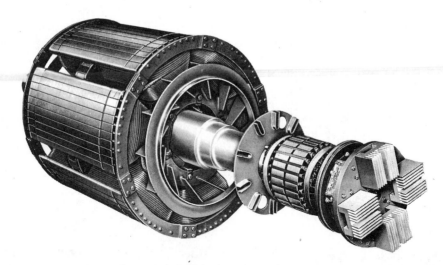

Fig. 20–7   Rotor of a brushless synchronous motor (*Photo courtesy of Electric Machinery, Turbodyne Division, Dresser Industries Inc.*)

Power Factor

A poor lagging power factor results when the field current is decreased below normal by inserting all of the resistance of the rheostat in the field circuit. The three-phase ac circuit to the stator supplies some magnetizing current which helps strengthen the weak dc field. This magnetizing component of current lags the voltage by 90 electrical degrees. Since the magnetizing component of current becomes a large part of the total current input, a low lagging power factor results.

If a weak dc field is strengthened, the power factor improves. As a result, the three-phase ac circuit to the stator supplies less magnetizing current. The magnetizing component of current becomes a smaller part of the total current input to the stator winding, and the power factor increases. If the field strength is increased sufficiently, the power factor increases to unity or 100 percent. When a power factor value of unity is reached, the three-phase ac circuit does not supply any current and the dc field circuit supplies all of the current necessary to maintain a strong rotor field. The value of dc field excitation required to achieve unity power factor is called *normal field excitation.*

If the magnetic field of the rotor is strengthened further by increasing the dc field current above the normal field excitation value, the power factor decreases. However, the power factor is leading when the dc field is overexcited. The three-phase ac circuit feeding the stator winding delivers a demagnetizing component of current which opposes the too strong rotor field. This action results in a weakening of the rotor field to its normal magnetic strength.

The diagrams in figure 20–8 show how the dc field is aided or opposed by the magnetic field set up by the ac windings. It is assumed in figure 20–8 that the dc field is stationary and a revolving armature is connected to the ac source. Keep in mind the fact that

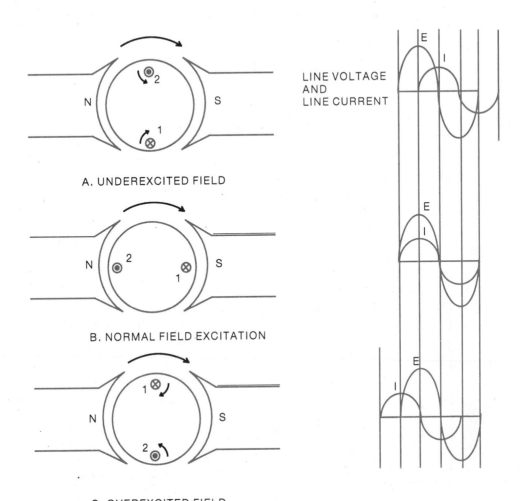

A. UNDEREXCITED FIELD

B. NORMAL FIELD EXCITATION

C. OVEREXCITED FIELD

LINE VOLTAGE
AND
LINE CURRENT

Fig. 20–8   Field excitation in a synchronous motor

most synchronous motors have stationary ac windings and a revolving dc field. For either case, however, the principle of operation is the same.

Figure 20–9 shows two characteristic operating curves for a three-phase synchronous motor. With normal full field excitation, the power factor has a peak value of unity or 100 percent and the ac stator current is at its lowest value. As the dc field current is decreased in value, the power factor decreases in the lag quadrant and there is a resulting rapid rise in the ac stator current. If the dc field current is increased above the normal field excitation value, the power factor decreases in the lead quadrant and a rapid rise in the ac stator current results.

It has been shown that a synchronous motor operated with an overexcited dc field has a leading power factor. For this reason, a three-phase synchronous motor often is connected to a three-phase industrial feeder circuit having a low lagging power factor. In

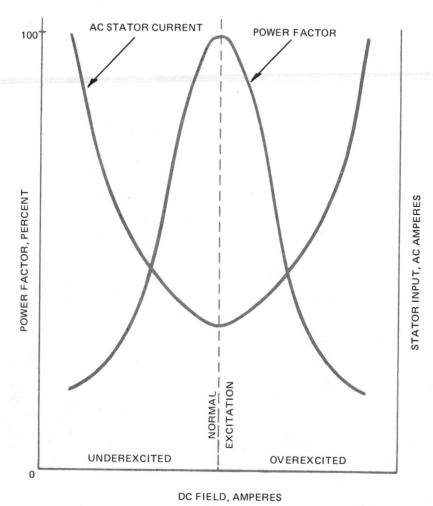

Fig. 20–9 Characteristic operating curves for synchronous motors

other words, the synchronous motor with an overexcited dc field will correct the power factor of the industrial feeder circuit.

In figure 20–10 two induction motors with lagging power factors are connected to an industrial feeder circuit. The synchronous motor connected to the same feeder is operated with an overexcited dc field. Since the synchronous motor can be adjusted so that the resulting power factor is leading, the power factor of the industrial feeder can be corrected until it reaches a value near unity or 100 percent.

Reversing Rotation

The direction of rotation of a synchronous motor is reversed by interchanging any two of the three line leads feeding the stator winding. The direction of rotation of the motor does not change if the two conductors of the dc source are interchanged.

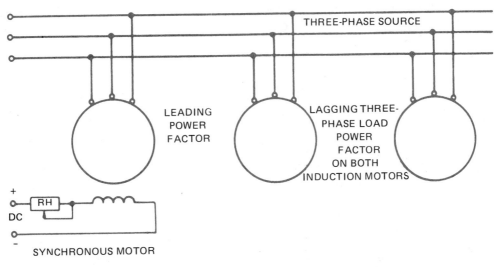

THREE-PHASE SOURCE

LEADING
POWER
FACTOR

LAGGING THREE-
PHASE LOAD
POWER
FACTOR
ON BOTH
INDUCTION MOTORS

+
RH
DC

–
SYNCHRONOUS MOTOR

Fig. 20–10   Synchronous motor used to correct power factor

INDUSTRIAL APPLICATIONS

The three-phase synchronous motor is used when a prime mover having a constant speed from a no-load condition to full load is required, such as fans, air compressors, and pumps. The synchronous motor is used in some industrial applications to drive a mechanical load and also correct power factor. In some applications, this type of motor is used only to correct the power factor of an industrial power system. When the synchronous motor is used only to correct power factor and does not drive any mechanical load, it serves the same purpose as a bank of capacitors used for power factor correction. Therefore, in such an installation the motor is called *asynchronous capacitor*.

Three-phase synchronous motors, up to a rating of 10 horsepower, usually are started directly across the rated three-phase voltage. Synchronous motors of larger sizes are started through a starting compensator or an automatic starter. In this type of starting, the voltage applied to the motor terminals at the instant of start is about half the value of the rated line voltage and the starting surge of current is limited.

SUMMARY

The ac synchronous motor is used where the speed must be kept constant. As the name implies, the motor will run at the designed synchronous speed. The principle used in the larger three-phase synchronous motors is to provide a dc field for the rotor. The methods may vary in the application of the dc. Some motors use an external dc source and feed the dc to the rotor via slip rings. Other motors will control a magnetic field to the rotor and use solid state rectifiers to create dc in the rotor. In either case, the rotor field can change the power factor of the synchronous motor and allow it to act as a source of leading-power factor, thereby correcting the normal lagging-power factor of an industrial power system.

ACHIEVEMENT REVIEW

A.   Completely answer the following questions.

1.   List the basic parts of a three-phase synchronous motor. _____

_____

_____

2.   What is an amortisseur winding?_____

_____

_____

3.   Explain the proper procedure to use in starting a synchronous motor. _____

_____

_____

_____

4.   A three-phase synchronous motor with six stator poles and six rotor poles is operated from a three-phase, 60-hertz line of the correct voltage rating. Determine the speed of the motor. _____

_____

_____

5.   How is a leading power factor obtained with a three-phase synchronous motor?

_____

_____

_____

6.   What is the purpose of a rheostat in the separately excited dc field circuit of a synchronous motor? _____

_____

_____

_____

7.   How is the direction of rotation of a three-phase synchronous motor reversed?

_____

_____

_____

8. State two important applications for three-phase synchronous motors.

_____

_____

9. What is a synchronous capacitor? _____

_____

B. Select the correct answer for each of the following statements and place the corresponding letter in the space provided.

10. A synchronous motor must be started   _____
    a. with full dc in the field circuit.
    b. with weak dc in the field circuit.
    c. as an induction motor.
    d. when the power factor is low.

11. The speed of a synchronous motor   _____
    a. is constant from no load to full load.
    b. drops from no load to full load.
    c. increases from no load to full load.
    d. is variable from no load to full load.

12. A synchronous motor with an underexcited dc field has   _____
    a. a leading power factor.
    b. a lagging power factor.
    c. less synchronous speed.
    d. no effect.

13. The power factor of a synchronous motor can be varied by
    changing the   _____
    a. brush polarity.
    b. phase rotation.
    c. speed of rotation.
    d. field excitation.

14. A synchronous motor running on the three-phase line voltage serves the same function of power factor correction as   _____
    a. bank of resistors.
    b. a bank of capacitors.
    c. an induction motor.
    d. a wound-rotor motor.

U•N•I•T
21
THREE-PHASE MOTOR INSTALLATIONS

OBJECTIVES

After studying this unit, the student will be able to
- determine, for several types of three-phase ac induction motors, the

    size of the conductors required for three-phase, three-wire branch circuits.

    sizes of fuses used to provide starting protection.

    disconnecting means required for the motor type.

    size of the thermal overload units required for running overcurrent protection.

    size of the main feeder to a motor installation.

    overcurrent protection required for the main feeder.

    main disconnecting means for the motor installation.
- use the National Electrical Code.

The work of the industrial electrician requires a knowledge of the National Electrical Code requirements which govern three-phase motor installations and the ability to apply these requirements to installations. The elements of a motor circuit are shown in figure 21–1.

This unit outlines the procedure for determining the wire size and the proper overload and starting protection for a typical three-phase motor installation. The motor installation example consists of a feeder circuit feeding three branch circuits. Each of the three branch circuits is connected to a three-phase motor of a specified horsepower rating. The feeder circuit and the branch circuits have the necessary overcurrent protection required by the National Electrical Code.

THREE-PHASE MOTOR LOAD

The industrial motor installation described in this example is connected to a 230-volt, three-phase, three-wire service (figure 21–2). The load of this system consists of the following branch circuits.

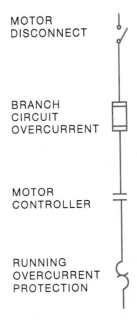

MOTOR DISCONNECT

BRANCH CIRCUIT OVERCURRENT

MOTOR CONTROLLER

RUNNING OVERCURRENT PROTECTION

Fig. 21–1 Line diagram of motor control system

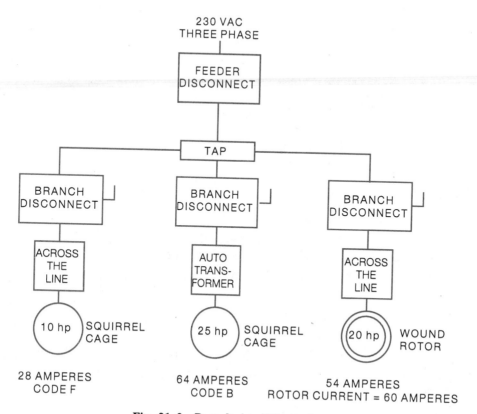

Fig. 21–2   Branch circuit for each motor

1. One branch circuit which feeds a three-phase, squirrel-cage induction motor rated at 230 volts, 28 amperes, 10 hp, with a code letter F marking.

2. One branch circuit which feeds a three-phase, squirrel-cage induction motor rated at 230 volts, 64 amperes, 25 hp, with a code letter B marking.

3. One branch circuit which feeds a three-phase, wound-rotor induction motor rated at 230 volts, 54 amperes, and 20 hp. The full-load rotor current is 60 amperes.

BRANCH CIRCUIT FOR EACH MOTOR

The values given in NEC *Tables 310-16, 310-17, 310-18 and 310-19*, including notes, shall be used with code book current for motors in determining ampacity of conductor and fuse size.

Three specific facts must be determined for each of the three branch circuits comprising the load of the installation.

1. The size of the conductors for each three-phase, three-wire branch circuit.

2. The fuse size to be used for short circuit protection. The fuses protect the wiring and the motor from any faults or short circuits in the wiring or motor windings.

3. The size of the thermal overload units to be used for running protection. The overload units protect the motor from potential damage due to a continued overload on the motor.

NOTE: The full-load amperes shall be taken from the motor's nameplate only for calculating thermal overload units (See NEC *Article 430-6*). Other calculations are based on code book rated values from 430-148, 149, 150.

BRANCH CIRCUIT 1

The first branch circuit feeds a three-phase, squirrel-cage induction motor. The nameplate data of this motor is as follows:

Squirrel-Cage Induction Motor	
Volts 230	Amperes 28
Phase 3	Speed 1,735 r/min
Code Letter F	Frequency 60 Hertz
10 Horsepower	Temperature Rating 40° Celsius

Conductor Size

Section 430-22(a) of the Code states that branch-circuit conductors supplying a single motor shall have a carrying capacity equal to not less than 125 percent of the fullload current rating of a motor. This general rule may be modified according to *Table 430-22(a) Exception* for certain special service classifications.

The following procedure is used to determine the size of the conductors of the branch circuit feeding the 10-hp motor.

a. The 10-hp motor has a full-load current rating of 28 amperes. According to *Section 430-152:*

$$28 \times 125\% = 35 \text{ amperes}$$

b. Using 35 amps and referring to *Table 310–16*, a proper size conductor is selected. This process requires the electrician to determine the temperature ratings of each termination used second the ampere rating of the eqipment circuit. According to NEC *Article 110–14(c)*, the temperature rating of the conductor, used to determine the ampere capacity(ampacity), must not exceed the temperature rating of any of the connections. Unless all the terminations are marked for a higher temperature, the column in 310–16 marked 60-degree-C is selected to determine the conductor ampacity. Even if using a standard building wire THHN, the conductor size is #8 in the 60-degree-C column.

c. Table C1 in NEC Appendix C indicates that 3 #8 THHN conductors will fit in a 1/2 inch conduit.

The squirrel-cage induction motor is to be connected directly across the rated line voltage through an across-the-line motor starter. The branch-circuit, short circuit and ground fault protection for this motor consists of three standard nontime-delay fuses enclosed in a safety switch located on the line side of the magnetic starter. According to *Section 430-109* of the Code, this switch shall be a motor-circuit switch with a horsepower rating, a circuit breaker, or a molded case switch and shall be a listed device. NOTE: The Underwriters' Laboratories, Inc. *Electrical Construction Materials List* states that "some enclosed switches have dual horsepower ratings, the larger of which is based on the use of fuses with time delay appropriate for the starting characteristics of the motor. Switches with such horsepower ratings are marked to indicate this limitation and are tested at the larger of the two ratings."

Motor Branch-Circuit Protection

The branch-circuit short circuit and ground fault protection for a three-phase, squirrel-cage induction motor marked with the code letter F is given in *Table 430-152*. For the branch circuit 1 motor being considered, the motor circuit overcurrent device shall not exceed 300 percent of the full-load current of the motor (nontime-delay fuses). *Article 430-52* with exceptions applies to *Table 430-152*.

The branch-circuit fuse protection for the branch circuit feeding the squirrel-cage motor is:

Since the 10-hp motor has a full-load current rating of 28 amperes, and given the appropriate value from *Table 430-152*:

$$28 \times 300\% = 84 \text{ amperes}$$

Section 430-52 states that if the values for branch-circuit protective devices as determined using the percentages in *Table 430-152* do not correspond to the standard device sizes or ratings, then the next larger size rating or setting should be used.

However, *Section 240-6* of the Code indicates that the next larger standard size fuse above 84 amperes is 90 amperes. Standard nontime-delay cartridge fuses rated at 90 amperes may be used as the branch-circuit protection for this motor circuit.

The branch circuit, short circuit, and ground fault protection may also be calculated using a time-delay fuse. Referring to *Table 430–152*, the second column is selected and 175% of 28 amps is calculated ($1.75 \times 28 = 49$ amps). The next larger size is used: in this example, 50 amp fuses would be the choice. The code does allow the electrician to increase the size of the fuse according to the exceptions to *430–52* c(1).

Three-Pole, Three-Fuse, 230-Volt Ac Safety Switches		
	Approximate Manufacturer Horsepower Ratings	
Amperes	Standard	Maximum
30	3	7 1/2 *
60	7 1/2	15 *
100	15	30 *
200	25	60 *
400	50	100 *

Fig. 21-3   Table for safety switches

Disconnecting Means

According to the table for safety switches (figure 21–3) the disconnecting means for this 10-hp motor is a 15-hp, 100-ampere safety switch in which the 90-ampere fuses are installed.

Since these safety switches are dual rated, it is permissible to install a 60-ampere safety switch having a maximum rating of 15 hp if the time-delay fuses are appropriate for the starting characteristics of the motor. The size of the time-delay fuses installed in the safety switch depends on the degree of protection desired and the type of service required of the motor. Time-delay fuses ranging in size from 35 amperes to 60 amperes may be installed in the safety switch.

Running Overcurrent Protection

The running overcurrent protection consists of three current monitors, usually thermal, housed in the across-the-line motor starter. (See the note following *Table 430-37 of the Code* for an exception to this statement.)

Section 430-32(a)(1) of the Code states that the running overcurrent protection (motor & branch circuit overload protection) for a motor shall trip at not more than 125 percent of the full-load current (as shown on the nameplate) for motors with a marked temperature rise not over 40 degrees Celsius.

The trip current of the thermal units used as running overcurrent protection is:

$$28 \times 125\% = 35 \text{ amperes}$$

When the selected overload relay is not sufficient to start the motor or to carry the load, *Section 430-34* permits the use of the next higher size or rating, but must trip at no more than 140 percent of the full-load motor current.

BRANCH CIRCUIT 2

A second branch circuit feeds a three-phase, squirrel-cage induction motor. The nameplate data for this motor is as follows:

Squirrel-Cage Induction Motor	
Volts 230	Amperes 64
Phase 3	Speed 1,740 r/min
Code Letter B	Frequency 60 Hertz
	25 Horsepower
	Temperature Rating 40° Celsius

Conductor Size

The following procedure is used to determine the size of the conductors of the branch circuit feeding the 25-horsepower motor.

a. The 25-hp motor has a full-load current rating of 68 amperes (see NEC *Table 430-150*). (According to Code *Section 430-22(a)*, 125% is needed for ampacity):

$$68 \times 125\% = 85 \text{ amperes}$$

b. *Table 310-16* indicates that a No. 3 Type TW or THHN copper conductor or a No. 3 Type THW conductor. (Assume 60° C terminations).

c. *Table C1 of Appendix C* shows that three No. 3 TW or THW conductors may be installed in a 1 1/4-inch conduit. A 1-inch conduit is required for three No. 3 THHN conductors.

NOTE: *Section 360-4F(c)* of the Code requires that where conductors of No. 4 size or larger enter an enclosure, an insulating bushing or equivalent must be installed on the conduit.

Motor Branch-Circuit Protection

The 25-hp squirrel-cage induction motor is to be started using an autotransformer. The branch-circuit overcurrent protection for this motor circuit consists of three non-time-delay fuses located in a safety switch mounted on the line side of the starting compensator.

For a squirrel-cage induction motor which is marked with code letter B and which is being used with a starting compensator, *Table 430-152* of the Code requires that the branch-circuit overcurrent protection not exceed 300 percent of the full-load current of the motor.

The branch-circuit overcurrent protection for the branch circuit feeding this motor is:

Since the 25-hp motor has a full-load current rating of 68 amperes (NEC *Table 430-150)*,

$$68 \times 300\% = 204 \text{ amperes}$$

Section 240-6 does not show 204 amperes as a standard size for a fuse. However, *Section 430-52* permits the use of a fuse of the next higher size if the calculated size is not a standard size. In this case, 200 amperes should be attempted. Therefore, three 200 ampere nontime-delay fuses can be used as the branch-circuit protection for this motor.

Disconnecting Means

According to the table for safety switches in figure 21-3, the disconnecting means for the 25-hp motor is a 25-hp, 200-ampere safety switch in which the 200 ampere fuses are installed.

Time-delay fuses may be installed in safety switches. In this example, 175% × 68A = 119A. 125 fuses are the next largest size and may be used according to exceptions to 430-52. The safty switch would be the same size.

Running Overcurrent Protection (Motor and Branch Circuit Overload Protection)

The running overcurrent protection consists of three magnetic overloads located in the starting compensator. According to the nameplate, the motor has a full-load current rating of 64 amperes. The current setting of the magnetic overload units is set to trip at

$$64 \times 125\% = 80 \text{ Amperes (trip current)}$$

BRANCH CIRCUIT 3

A third branch circuit feeds a three-phase, wound-rotor induction motor. The nameplate data for this motor is as follows:

Wound-Rotor Induction Motor	
Volts 230	Stator Amperes 54
Phase 3	Rotor Amperes 60
Frequency 60 Hertz	20 Horsepower
Temperature Rating 40° Celsius	

Conductor Size (Stator)

The following procedure is used to determine the size of the conductors of the branch circuit feeding the 20-horsepower motor.

    a. The 20-hp motor has a full-load current rating of 54 amperes. According to NEC *Section 430-22(a), and Table 430-150,*

$$54 \times 125\% = 67.5 \text{ amperes}$$

    b. *Table 310-16* indicates that a No. 4 Type TW, THW, THHN conductor (70 amperes).

    c. *Tables C1 of Appendix C* show that three No. 4 TW or THW, or THHN conductors may be installed in a 1-inch conduit.

NOTE: *Article 300-4F(c)* requires that where conductors of No. 4 size or larger enter an enclosure, an insulating bushing or equivalent must be installed on the conduit.

Motor Branch-Circuit Protection

The 20-hp wound-rotor induction motor is to be started by means of an across-the-line magnetic motor switch. This motor starter applies the rated three-phase voltage to the stator winding. Speed control is provided by a manual drum controller used in the rotor or secondary circuit. All of the resistance of the controller is inserted in the rotor circuit when the motor is started. As a result, the inrush starting current to the motor is less than if the motor were started at the full voltage.

The branch-circuit overcurrent protection of a wound-rotor induction motor is required by *Table 430-152* of the Code not to exceed 150 percent of the full-load running current of the motor.

The branch-circuit overcurrent protection for the branch circuit feeding this motor is: The full-load current equals 54 amperes for a 20-hp wound-rotor motor

$$54 \times 150\% = 81 \text{ amperes}$$

Section 240-6 does not show 81 amperes as a standard fuse size. Article *430-52* allows the next larger size. A 90A fuse should be chosen.

Disconnecting Means

According to the table for safety switches in figure 14-3, the disconnecting means for the 20 hp motor is a 25-hp, 200-ampere safety switch. Reducers must be installed in this switch to accommodate the 90-ampere fuses required for the motor branch circuit protection. Because of the dual rating of these safety switches, it is permissible to use a 100-ampere switch having a maximum rating of 30 hp. In this case, standard 90-ampere nontime-delay fuses or 90-ampere time-delay fuses may be installed.

Running Overcurrent Protection (Motor Overload Protection)

The running overcurrent protection consists of three thermal overload units located in the across-the-line magnetic motor starter (except as indicated in the note following *Table 430-37*). According to the nameplate, the motor has a full-load current rating of 54 amperes. The rated trip current of each thermal unit is:

$$54 \times 125\% = 67.5 \text{ amperes}$$

Conductor Size (Rotor)

The rotor winding of the 20-hp, wound-rotor induction motor is rated at 60 amperes. The following procedure is used to determine the size of the conductors for the secondary circuit from the rotor slip rings to the drum controller.

a. *Section 430-23(a)* requires that the conductors connecting the secondary of a wound-rotor induction motor to its controller have a current-carrying capacity not less than 125 percent of the full-load secondary current of the motor for continuous duty.

$$60 \times 125\% = 75 \text{ amperes}$$

b. *Table 310-16* indicates that several types of copper conductors can be used: No. 3 Type TW, Type THW, or Type THHN, asssuming 60° terminations.

c. *Table C1 of Appendix C* shows that three No. 3 TW conductors can be installed in a 1¼ inch conduit. A 1¼-inch conduit is required if three No. 3 THW conductors are used. A 1-inch conduit is required for three No. 3 THHN wires.

NOTE: *Article 300-4F(c)* requires the use of insulating bushings or equivalent on all conduits containing conductors of No. 4 size or larger entering enclosures.

If the resistors are mounted outside the speed controller, the current capacity of the conductors between the controller and the resistors shall be not less than the values given in *Table 430-23(c)*.

For example, the manual speed controller used with the 20-hp wound-rotor induction motor is to be used for heavy intermittent duty. *Section 430-23(c)* requires that the conductors connecting the resistors to the speed controller have an ampacity not less than 85 percent of rated rotor current.

$$60 \times 85\% = 51 \text{ amperes}$$

Table 310-16 indicates that 51 amperes can be carried safely by No. 6 wire. As a result, the temperatures generated at the resistor location are an important consideration.

Section 430-32(d) states that the secondary circuits of wound-rotor induction motors, including the conductors, controllers and resistors, shall be considered as protected against overload by the motor running overcurrent protection in the primary or stator circuits, Therefore, no overcurrent protection is necessary in the secondary rotor circuit.

MAIN FEEDER

When the conductors of a feeder supply two or more motors, the required wire size is determined using Code rules. *Section 430-24* of the Code states that the feeder shall have an ampacity of not less than 125 percent of the full-load current of the highest rated motor of the group plus the sum of the full-load current ratings of the remaining motors in the group. The full-load current of the motor is taken from NEC *Table 430-150.*

The motor with the largest full-load running current is the 25-hp motor. This motor has a full-load current rating of 68 amperes. The main feeder size, then, in compliance with *Section 430-24,* is:

$$68 \times 125\% = 85 \text{ amperes}$$

Then: $85 + 54 + 28 = 167$ amperes.

Table 310-16 indicates that No. 4/0 Type TW or Type THHN copper conductors can be used when using 60° terminations.

Table C1 of Appendix C show that three No. 4/0 TW conductors can be installed in 2-inch conduit. Three No. 4/0 THHN conductors can be installed in a 2-inch conduit.

Main Feeder Short-Circuit Protection

Section 430-62(a) states that a feeder which supplies motors shall be provided with overcurrent protection. The feeder overcurrent protection shall not be greater than the largest current rating of the branch-circuit protective device for any motor of the group, based on *Table 430-152,* plus the sum of the full-load currents of the other motors of the group.

The branch circuit feeding the 25-hp motor has the largest value of overcurrent protection. This value, as determined from *Table 430-152,* is 170 amperes (68×300 or 200 amperes.)

The full-load current rating of the 20-hp motor is 54 amperes, and the full-load current rating of the 10-hp motor is 28 amperes. The size of the fuses to be installed in the main feeder cicuit shall not be greater than the sum of $200 + 54 + 28 = 282$ amperes.

Therefore, three 250-ampere nontime-delay fuses are used for the feeder circuit. This procedure should be in conformance with *Example 8, Chapter* 9 of the Code. Exceptions may be made if the fuses do not allow the motor to start or run.

Main Disconnecting Means

Section 430-109 lists several exceptions to the ruling that the disconnecting means shall be a motor-circuit switch, rated in horsepower, or a circuit breaker. The disconnecting means shall have a carrying capacity of at least 115 percent of the sum of the current ratings of the motors, *Section 430-110 (c1 and 2).* Therefore, the 250 ampere fuses required as the overcurrent protection for the main feeder are installed in a 400-ampere safety switch.

Wire types and sizes are selected by the ambient temperatures of the place of installation and the economics of the total installation, such as the minimum size conduits, cost of the wire sizes, and the cost of the labor to install the different selections.

SUMMARY

The motor installation is one of the hardest calculations to perform and get all the components correct, in the proper location, and at the correct size. The code book guides you through the main components of the calculation but you must know where to look and how to apply the proper codes. There are many facets to the correct installation including: feeder and feeder protection, branch circuit and branch protection, conductor sizes and overcurrent protection, running overcurrent protection and secondary circuit protection.

ACHIEVEMENT REVIEW

A feeder circuit feeds three branch motor circuits. Branch motor circuit No. 1 has a load consisting of an induction motor with the following nameplate data:

No. 1

Squirrel-Cage Induction Motor	
230 Volts	15 Amperes
3 Phase	60 Hertz
5 Horsepower	Code Classification D
Temperature Rating 40° Celsius	

Branch motor circuit No. 2 has a load consisting of an induction motor with the following nameplate data: (This motor is equipped with an autotransformer starting compensator.):

No. 2

Squirrel-Cage Induction Motor	
230 Volts	40 Amperes
3 Phase	60 Hertz
15 Horsepower	Code Classification F
Temperature Rating 40° Celsius	

Branch motor circuit No. 3 has a load consisting of a wound-rotor induction motor with the following nameplate data:

Wound-Rotor Induction Motor	
230 Volts	22 Stator Amperes
3 Phase	26 Rotor Amperes
7 1/2 Horsepower	60 Hertz
Continuous Duty	
Temperature Rating 40° Celsius	

No. 3

1. Refer to the following diagram.
   a. Determine the running overload protection in amperes required for the motor in branch circuit No. 1.
   b. Determine the appropriate wire size (TW).
      (Insert the answers on the diagram.)

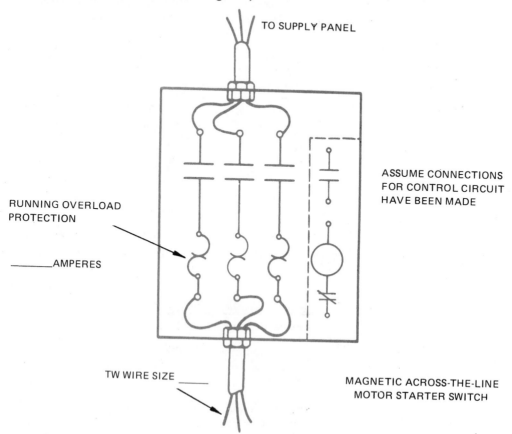

TO SUPPLY PANEL

ASSUME CONNECTIONS
FOR CONTROL CIRCUIT
HAVE BEEN MADE

RUNNING OVERLOAD
PROTECTION

_____AMPERES

TW WIRE SIZE _____

MAGNETIC ACROSS-THE-LINE
MOTOR STARTER SWITCH

2. Refer to the following diagram.
   a. Determine the running overload protection in amperes required for the motor in branch circuit No. 2.
   b. Determine the appropriate wire size of the copper TW conductors. Note that the 15-hp squirrel-cage induction motor in this circuit is started by means of a starting compensator.
   (Insert the answers on the diagram.)

TO SUPPLY PANEL

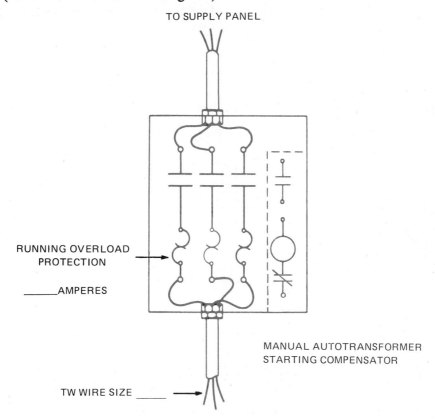

RUNNING OVERLOAD
PROTECTION

_____AMPERES

MANUAL AUTOTRANSFORMER
STARTING COMPENSATOR

TW WIRE SIZE _____

3. Refer to the following diagram.
   a. Determine the running overload protection in amperes required for the motor in branch circuit No. 3.
   b. Determine the appropriate wire size of the copper conductors.
      (Insert the answers on the diagram.)
   c. Determine the size of the conductors required for the secondary circuit of the wound-rotor induction motor in branch circuit No. 3. The secondary or rotor circuit feeds between the slip rings of the wound rotor and the speed controller. Indicate the size of the conduit. Use TW conductors.

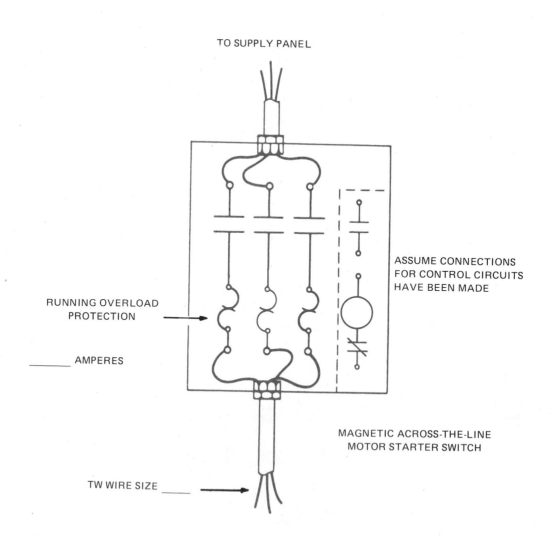

TO SUPPLY PANEL

ASSUME CONNECTIONS
FOR CONTROL CIRCUITS
HAVE BEEN MADE

RUNNING OVERLOAD
PROTECTION

_____ AMPERES

MAGNETIC ACROSS-THE-LINE
MOTOR STARTER SWITCH

TW WIRE SIZE _____

4. Refer to the following diagram.
   a. Determine the current rating in amperes of the fuses (nontime-delay) used as overload protection for the main feeder circuit shown in the diagram.
   b. Determine the TW conductor size for the main feeder switch.
      (Insert the answers on the diagram.)

TO SERVICE ENTRANCE

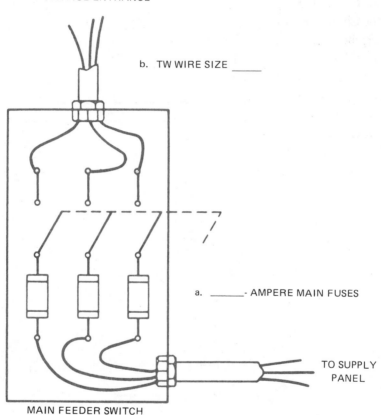

b. TW WIRE SIZE _____

a. _____ - AMPERE MAIN FUSES

TO SUPPLY
PANEL

MAIN FEEDER SWITCH

5. Refer to the following diagram.
    a. Using TW-type copper conductors, determine the size of the conductors and
       conduit required for the main feeder circuit which feeds the three branch motor
       circuits. Indicate the sizes on the diagram.
    b. Determine the size of fuses in amperes required for the starting overload
       protection for each of the branch circuits.
       Motor Circuit No. 1 _____
       Motor Circuit No. 2 _____
       Motor Circuit No. 3 _____
       (Insert the answers on the diagram.)
    c. Using TW-type copper conductors, determine the size of rigid conduit required
       for each of the three branch circuits.
       Motor Circuit No. 1 _____
       Motor Circuit No. 2 _____
       Motor Circuit No. 3 _____
    (Insert the answers on the diagram.)

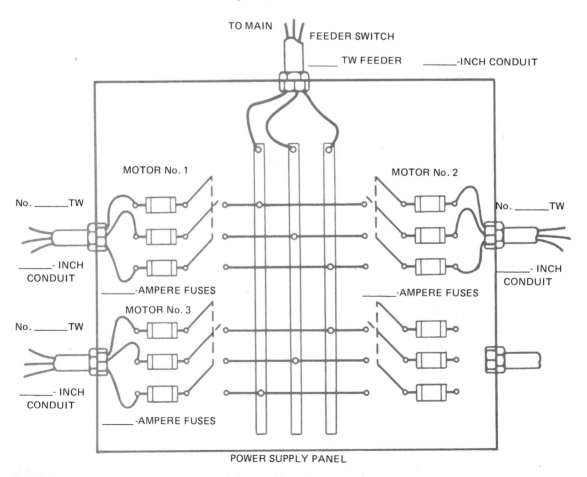

U•N•I•T
22

SINGLE-PHASE INDUCTION MOTORS

OBJECTIVES

After studying this unit, the student will be able to

- describe the basic operation of the following types of induction motors:
  split-phase motor (both single and dual voltage)
  capacitor start, induction run motor (both single and dual voltage)
  capacitor start, capacitor run motor with one capacitor
  capacitor start, capacitor run motor with two capacitors
  capacitor start, capacitor run motor having an autotransformer with one capacitor
- compare the motors in the listing of objective 1 with regard to starting torque, speed performance, and power factor at the rated load.

The two principal types of single-phase induction motors are the split-phase motor and the capacitor motor. Both types of single-phase induction motors usually have a fractional horsepower rating. The split-phase motor is used to operate such devices as washing machines, small water pumps, oil burners, and other types of small loads not requiring a strong starting torque. The capacitor motor generally is used with devices requiring a strong starting torque, such as refrigerators and compressors. Both types of single-phase induction motors are relatively low in cost, have a rugged construction, and exhibit a good operating performance.

CONSTRUCTION OF A SPLIT-PHASE INDUCTION MOTOR

The split-phase induction motor basically consists of a stator, a rotor, a centrifugal switch located inside the motor, two end shields housing the bearings that support the rotor shaft, and a cast steel frame into which the stator core is pressed. The two end shields are bolted to the cast steel frame. The bearings housed in the end shields keep the rotor centered within the stator so that it will rotate with a minimum of friction and without striking or rubbing the stator core.

The stator for a split-phase motor consists of two windings held in place in the slots of a laminated steel core. The two windings consist of insulated coils distributed and connected to make up two windings spaced 90 electrical degrees apart. One winding is the running winding and the second winding is the starting winding.

The running winding consists of insulated copper wire. It is placed at the bottom of the stator slots. The wire size in the starting winding is smaller than that of the running winding, These coils are placed on top of the running winding coils in the stator slots nearest to the rotor.

Both the starting and running windings are connected in parallel to the single-phase line when the motor is started. After the motor accelerates to a speed equal to approximately two-thirds to three-quarters of the rated speed, the starting winding is disconnected automatically from the line by means of a centrifugal switch.

The rotor for the split-phase motor has the same construction as that of a three-phase, squirrel-cage induction motor. That is, the rotor consists of a cylindrical core assembled from steel laminations. Copper bars are mounted near the surface of the rotor. The bars are brazed or welded to two copper end rings. In some motors, the rotor is a one-piece cast aluminum unit.

Figure 22–1 shows a typical squirrel-cage rotor for a single-phase induction motor. This type of rotor requires little maintenance since there are no windings, brushes, slip rings, or commutators. Note in the figure that the rotor fans are a part of the squirrel-cage rotor assembly. These rotor fans maintain air circulation through the motor to prevent a large increase in the temperature of the windings.

The centrifugal switch is mounted inside the motor. The centrifugal switch disconnects the starting winding after the rotor reaches a predetermined speed, usually two-thirds to three-quarters of the rated speed. The switch consists of a stationary part and a rotating part. The stationary part is mounted on one of the end shields and has two contacts which act like a single-pole, single-throw switch. The rotating part of the centrifugal switch is mounted on the rotor.

Fig. 22–1   Cast aluminum squirrel-cage rotor (*Photo courtesy of General Electric Company*)

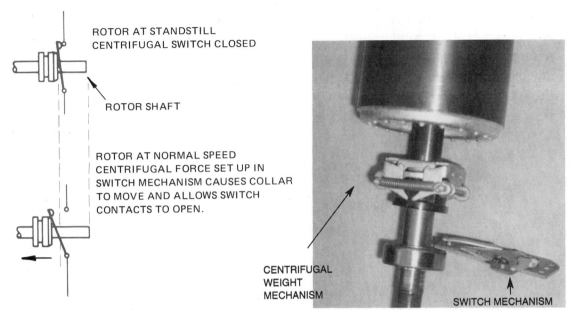

ROTOR AT STANDSTILL
CENTRIFUGAL SWITCH CLOSED

ROTOR SHAFT

ROTOR AT NORMAL SPEED
CENTRIFUGAL FORCE SET UP IN
SWITCH MECHANISM CAUSES COLLAR
TO MOVE AND ALLOWS SWITCH
CONTACTS TO OPEN.

CENTRIFUGAL
WEIGHT
MECHANISM

SWITCH MECHANISM

Fig. 22–2 Diagram shows the operation of a centrifugal switch

Fig. 22–3 Centrifugal switch mechanism with switch removed

A simple diagram of the operation of a centrifugal switch is given in figure 22–2. When the rotor is at a standstill, the pressure of the spring on the fiber ring of the rotating part keeps the contacts closed. When the rotor reaches approximately three-quarters of its rated speed, the centrifugal action of the rotor causes the spring to release its pressure on the fiber ring and the contacts open. As a result, the starting winding circuit is disconnected from the line. Figure 22–3 is a typical centrifugal switch used with split-phase induction motors.

Principle of Operation

When the circuit to the split-phase induction motor is closed, both the starting and running windings are energized in parallel. Because the running winding consists of a relatively large size of wire, its resistance is low. Recall that the running winding is placed at the bottom of the slots of the stator core. As a result the inductive reactance of this winding is comparatively high due to the mass of iron surrounding it. Since the running winding has a low resistance and a high inductive reactance, the current of the running winding lags behind the voltage approximately 90 electrical degrees.

The starting winding consists of a smaller size of wire; therefore, its resistance is high. Since the winding is placed near the top of the stator slots, the mass of iron surrounding it is comparatively small and the inductive reactance is low. Therefore, the starting winding has a high resistance and a low inductive reactance. As a result, the current of the starting winding is nearly in phase with the voltage.

The current of the running winding lags the current of the starting winding by approximately 30 electrical degrees. These two currents spaced 30 electrical degrees apart pass through these windings and a rotating magnetic field is developed. This field travels around the inside of the stator core. The speed of the magnetic field is determined using the same procedure given for a three-phase induction motor.

If a split-phase induction motor has four poles on the stator windings and is connected to a single-phase, 60-Hz source, the synchronous speed of the revolving field is:

$$S = \frac{120 \times f}{4} \qquad \begin{array}{l} S = \text{synchronous speed} \\ f = \text{frequency in hertz} \end{array}$$

$$S = \frac{120 \times 60}{4}$$

$$S = 1,800 \text{ r/min}$$

As the rotating stator field travels at the synchronous speed, it cuts the copper bars of the rotor and induces voltages in the bars of the squirrel-cage winding. These induced voltages set up currents in the rotor bars. As a result, a rotor field is created which reacts with the stator field to develop the torque which causes the rotor to turn.

As the rotor accelerates to the rated speed, the centrifugal switch disconnects the starting winding from the line. The motor then continues to operate using only the running winding. Figure 22–4 illustrates the connections of the centrifugal switch at the instant the motor starts (switch closed) and when the motor reaches its normal running speed (switch open).

A split-phase motor must have both the starting and running windings energized when the motor is started. The motor resembles a two-phase induction motor in which the currents of these two windings are approximately 90 electrical degrees out of phase. The voltage source, however, is single-phase; therefore, the motor is called a split-phase motor because it starts like a two-phase motor from a single-phase line. Once the motor accelerates to a value near its rated speed, it operates on the running winding as a single-phase induction motor.

If the centrifugal switch contacts fail to close when the motor stops, then the starting winding circuit is still open. When the motor circuit is reenergized, the motor will not start. The motor must have both the starting and running windings energized at the instant the motor circuit is closed to create the necessary starting torque. If the motor does not start but simply gives a low humming sound, then the starting winding circuit is open. Either the centrifugal switch contacts are not closed, or there is a break in the coils of the starting windings. **This is an unsafe condition**. The running winding will draw excessive current and, therefore, the motor must be removed from the line supply.

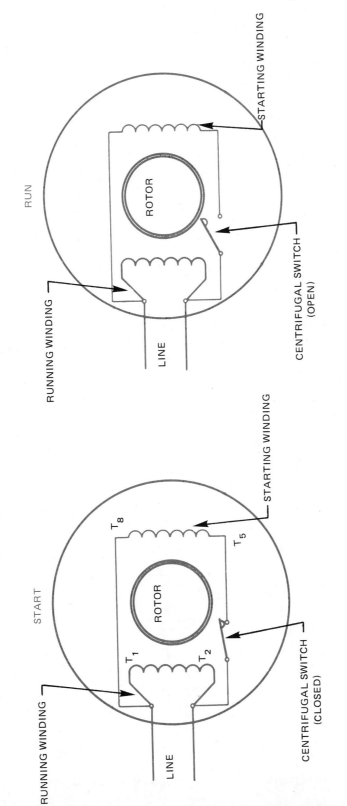

SPLIT-PHASE INDUCTION MOTOR

THE CENTRIFUGAL SWITCH OPENS AT APPROXIMATELY 75 PERCENT OF RATED SPEED

THE STARTING WINDING HAS HIGH RESISTANCE AND LOW INDUCTIVE REACTANCE.
THE RUNNING WINDING HAS LOW RESISTANCE AND HIGH INDUCTIVE REACTANCE.
(PRODUCES 45°–50° PHASE ANGLE FOR STARTING TORQUE.)

Fig. 22–4   Connections of the centrifugal switch at start and at run

If the mechanical load is too great when a split-phase motor is started, or if the terminal voltage applied to the motor is low, then the motor may fail to reach the speed required to operate the centrifugal switch.

The starting winding is designed to operate across the line voltage for a period of only three or four seconds while the motor is accelerating to its rated speed. It is important that the starting winding be disconnected from the line by the centrifugal switch as soon as the motor accelerates to 75 percent of the rated speed. Operation of the motor on its starting winding for more than 60 seconds may burn the insulation on the winding or cause the winding to burn out.

To reverse the rotation of the motor, simply interchange the leads of the starting winding (figure 22–5). This causes the direction of the field set up by the stator windings to become reversed. As a result, the direction of rotation is reversed. The direction of rotation of the split-phase motor can also be reversed by interchanging the two running winding leads. Normally, the starting winding is used for reversing.

Single-phase motors often have dual-voltage ratings of 115 volts and 230 volts. To obtain these ratings the running winding consists of two sections. Each section of the winding is rated at 115 volts. One section of the running winding is generally marked T_1 and T_2 and the other section is marked T_3 and T_4. If the motor is to be operated on 230 volts, the two 115-volt windings are connected in series across the 230-volt line. If the motor is to be operated on 115 volts, then the two 115-volt windings are connected in parallel across the 115-volt line.

The starting winding, usually consists of only one 115-volt winding. The leads of the starting winding are generally marked T_5 and T_8. If the motor is to be operated on 115

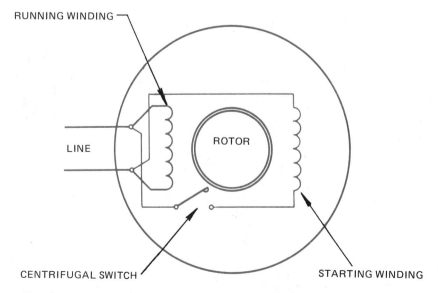

Fig. 22–5   Reversing direction of rotation on split-phase induction motor

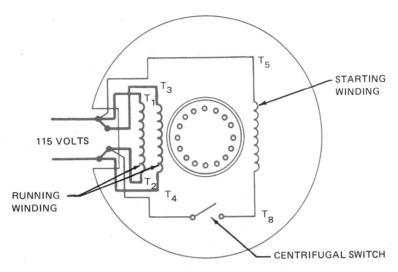

Fig. 22–6 Dual-voltage motor connected for 115 volts

volts, both sections of the running winding are connected in parallel with the starting winding (figure 22–6).

For 230-volt operation, the connection jumpers are changed in the terminal box so that the two 115-volt sections of the running winding are connected in series across the 230-volt line (figure 22–7). Note that the 115 volt starting winding is connected in parallel with one section of the running winding. The voltage drop across this section of the running winding is 115 volts, and the voltage across the starting winding is also 115 volts.

Some dual-voltage, split-phase motors have a starting winding with two sections as well as a running winding with two sections. The running winding sections are marked T_1 and T_2 for one section and T_3 and T_4 for the other section. One section of the starting winding is marked T_5 and T_6 and the second section of the starting winding is marked T_7 and T_8.

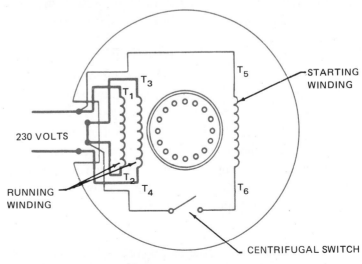

Fig. 22–7 Dual-voltage
motor connected for 230 volts

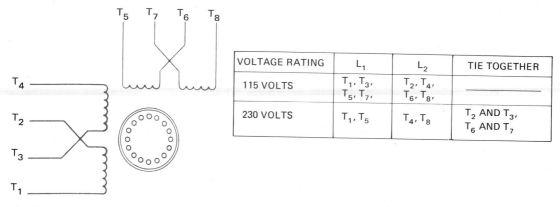

Fig. 22–8 Winding arrangement for dual-voltage motor with two starting and two running windings

The National Electrical Manufacturers Association (NEMA) has color coded the terminal leads. If colors are used, they should be coded as follows: T_1 – blue; T_2 – white; T_3 – orange; T_4 – yellow; T_5 – black; and T_6 – red.

Figure 22–7 shows the winding arrangement for a dual-voltage motor with two starting windings and two running windings. The correct connections for 115-volt operation and for 230-volt operation are given in the table shown in figure 22–8.

The speed regulation of a split-phase induction motor is very good. It has a speed performance from no load to full load that is similar to that of a three-phase, squirrel-cage induction motor. The percent slip on most fractional horsepower split-phase motors is from 4 percent to 6 percent.

The starting torque of the split-phase motor is comparatively poor. The low resistance and high inductive reactance in the running winding circuit, and the high resistance and low inductive reactance in the starting winding circuit cause the two current values to be considerably less than 90 electrical degrees apart. The currents of the starting and running windings in many split-phase motors are only 30 electrical degrees out of phase with each other. As a result, the field set up by these currents does not develop a strong starting torque.

CAPACITOR START, INDUCTION RUN MOTOR

The construction of a capacitor start motor is nearly the same as that of a split-phase induction motor. For the capacitor start motor, however, a capacitor is connected in series with the starting windings. The capacitor usually is mounted in a metal casing on top of the motor. The capacitor may be mounted in any convenient external position on the motor frame and, in some cases, may be mounted inside the motor housing. The capacitor provides a higher starting torque than is obtainable with the standard split-phase motor. In addition, the capacitor limits the starting surge of current to a lower value than is developed by the standard split-phase motor.

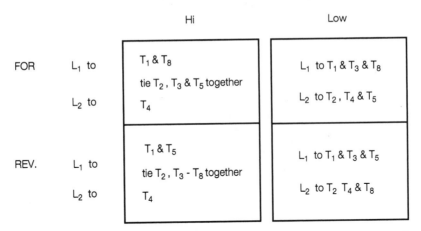

		Hi	Low
FOR	L_1 to	T_1 & T_8 tie T_2, T_3 & T_5 together	L_1 to T_1 & T_3 & T_8
	L_2 to	T_4	L_2 to T_2, T_4 & T_5
REV.	L_1 to	T_1 & T_5 tie T_2, T_3 - T_8 together	L_1 to T_1 & T_3 & T_5
	L_2 to	T_4	L_2 to T_2 T_4 & T_8

Fig. 22–9   Two running winding connection and one start winding connection chart

The capacitor start induction motor is used on refrigeration units, compressors, oil burners, and for small machine equipment, as well as for applications which require a strong starting torque.

Principle of Operation

When the capacitor start motor is connected for lower voltage and is started, both the running and starting windings are connected in parallel across the line voltage as the centrifugal switch is closed. The starting winding, however, is connected in series with the capacitor. When the motor reaches a value of 75 percent of its rated speed, the centrifugal switch opens and disconnects the starting winding and the capacitor from the line. The motor then operates as a single-phase induction motor using only the running winding. The capacitor is used to improve the starting torque and does not improve the power factor of the motor.

To produce the necessary starting torque, a revolving magnetic field must be set up by the stator windings. The starting winding current will lead the running winding current by 90 electrical degrees if a capacitor having the correct capacity is connected in series with the starting winding. As a result, the magnetic field developed by the stator windings is almost identical with that of a two-phase induction motor. The starting torque for a capacitor start motor thus is much better than that of a standard split-phase motor.

Defective capacitors are a frequent cause of malfunctions in capacitor start, induction run motors. Some capacitor failures that can occur are:

- the capacitor may short itself out, as evidenced by a lower starting torque.
- the capacitor may be "opened," in which case the starting winding circuits will be open, causing the motor to fail to start.
- the capacitor may short circuit and cause the fuse protection for the branch motor circuit to blow. If the fuse ratings are quite high and do not interrupt the power supply to the motor soon enough, the starting winding may burn out.

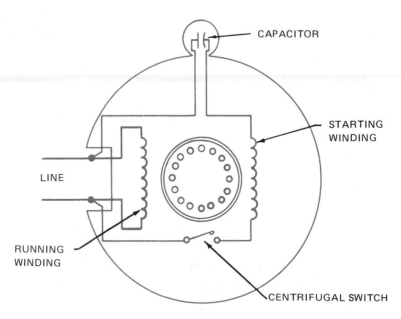

CAPACITOR

STARTING
WINDING

LINE

RUNNING
WINDING

CENTRIFUGAL SWITCH

Fig. 22–10   Connections for a capacitor start, induction motor

- starting capacitors can short circuit if the motor is turned on and off many times in a short period of time. To prevent capacitor failures, many motor manufacturers recommend that a capacitor start motor be started no more than 20 times per hour. Therefore, this type of motor is used only in those applications where there are relatively few starts in a short time period.

The speed performance of a capacitor start motor is very good. The increase in percent slip from a no-load condition to full load is from 4 percent to 6 percent. The speed performance then is the same as that of a standard split-phase motor.

The leads of the starting winding circuit are interchanged to reverse the direction of rotation of a capacitor start motor. As a result, the direction of rotation of the magnetic field developed by stator windings reverses in the stator core, and the rotation of the rotor is reversed. (See figure 22–9 for reversing lead connections.)

Figure 22–10 is a diagram of the circuit connections of a capacitor start motor before the starting winding leads are interchanged to reverse the direction of rotation of the rotor. The diagram in figure 22–11 shows the circuit connections of the motor after the starting winding leads are interchanged to reverse the direction of rotation.

A second method of changing the direction of rotation of a capacitor start motor is to interchange the two running winding leads. However, this method is seldom used.

Capacitor start, induction run motors often have dual-voltage ratings of 115 volts and 230 volts. The connections for a capacitor start motor are the same as those for split-phase induction motors.

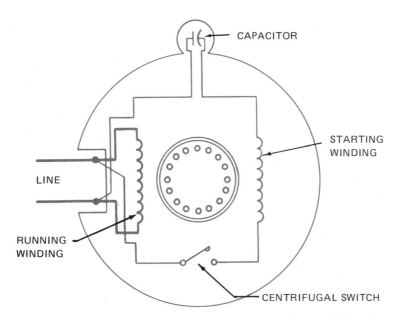

CAPACITOR

STARTING
WINDING

LINE

RUNNING
WINDING

CENTRIFUGAL SWITCH

Fig. 22–11 Connections for reversing a capacitor start, induction run motor

CAPACITOR START, CAPACITOR RUN MOTOR

The capacitor start, capacitor run motor is similar to the capacitor start, induction run motor, except that the starting winding and capacitor are connected in the circuit at all times. This motor has a very good starting torque. The power factor at the rated load is nearly 100 percent or unity due to the fact that a capacitor is used in the motor at all times.

There are several different designs for this type of motor. One type of capacitor start, capacitor run motor has two stator windings which are spaced 90 electrical degrees apart. The main or running winding is connected directly across the rated line voltage. A capacitor is connected in series with the starting winding and this series combination also is connected across the rated line voltage. A centrifugal switch is not used because the starting winding is energized through the entire operating period of the motor.

Figure 22–12 illustrates the internal connections for a capacitor start, capacitor run motor using one value of capacitance.

To reverse the rotation of this motor, the lead connections of the starting winding must be interchanged. This type of capacitor start, capacitor run motor is quiet in operation and is used on oil burners, fans, and small wood working and metal working machines.

A second type of capacitor start, capacitor run motor has two capacitors. Figure 22–13 is a diagram of the internal connections of the motor. At the instant the motor is started, the two capacitors are in parallel. When the motor reaches 75 percent of the rated speed, the centrifugal switch disconnects the larger capacity capacitor. The motor then operates with the smaller capacitor only connected in series with the starting winding.

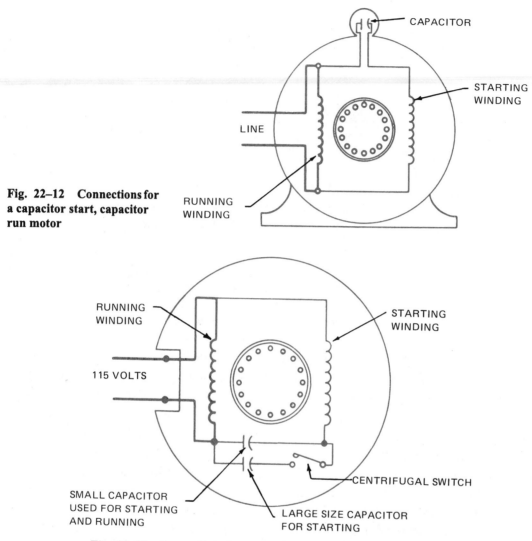

Fig. 22–12 Connections for a capacitor start, capacitor run motor

CAPACITOR

STARTING WINDING

LINE

RUNNING WINDING

RUNNING WINDING

STARTING WINDING

115 VOLTS

CENTRIFUGAL SWITCH

SMALL CAPACITOR USED FOR STARTING AND RUNNING

LARGE SIZE CAPACITOR FOR STARTING

Fig. 22–13   Connections for a capacitor start, capacitor run motor

This type of motor has a very good starting torque, good speed regulation, and a power factor of nearly 100 percent at rated load. Applications for this type of motor include furnace stokers, refrigerator units, and compressors.

A third type of capacitor start, capacitor run motor has an autotransformer with one capacitor. This motor has a high starting torque and a high operating power factor. Figure 22–14 is a diagram of the internal connections for this motor. When the motor is started, the centrifugal switch connects winding 2 to point A on the tapped autotransformer. As the capacitor is connected across the maximum transformer turns, it receives maximum voltage output on start. The capacitor thus is connected across a value of approximately

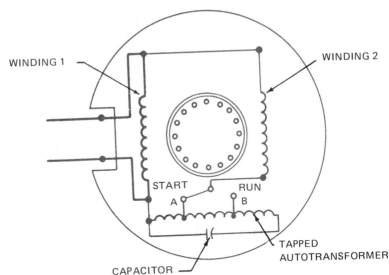

Fig. 22–14 Connections for a capacitor start, capacitor run motor with autotransformer

500 volts. As a result, there is a high value of leading current in winding 2 and a strong starting torque is developed.

When the motor reaches approximately 75 percent of the rated speed, the centrifugal switch disconnects the starting winding from point A and reconnects this winding to point B on the autotransformer. Less voltage is applied to the capacitor, but the motor operates with both windings energized. Thus, the capacitor maintains a power factor near unity at the rated load.

The starting torque of this motor is very good and the speed regulation is satisfactory. Applications requiring these characteristics include large refrigerators and compressors.

NATIONAL ELECTRICAL CODE REGULATIONS

Section 430-32(b)(1) of the National Electrical Code states that any motor of one horsepower or less which is manually started and is within sight from the starter location, shall be considered as protected against overload by the overcurrent device protecting the conductors of the branch circuit. This branch overcurrent device shall not be larger than specified in *Article 430, Part D (Motor Branch Circuit, Short-Circuit and Ground-Fault Protection)*. An exception is that any such motor may be used at 120 volts or less on a branch circuit protected at not over 20 amperes.

A distance of more than 50 feet is considered to be out of sight from the starter location. *Section 430-32(c)* covers motors, one horsepower or less, automatically started, which are out of sight from the starter location or permanently installed.

Section 430-32(c)(1) states that any motor of one horsepower or less which is started automatically shall have a separate overcurrent device which is responsive to the motor current. This overload unit shall be set to trip at not more than 125 percent of the full-load current rating of the motor for motors marked to have a temperature rise not over 40 degrees Celsius or with a service factor not less than 1.15, (1.15 or higher) and not more than 115 percent for all other types of motors.

SUMMARY

The single-phase induction motor is one of the most used residential and light commercial motors. Each application will dictate the correct motor style to use. All the motors use the concept of taking one phase, or one sinewave, and shifting the effects of the currents through the coils to create a moving magnetic field. The split-phase and the capacitor-start motor utilize a starting switch to disconnect the starting windings from the line once the motor is up to running speed. Two-capacitor motors use multiple capacitors or variations of two value capacitors to create a starting and a running circuit. All the same NEC regulations that apply to three-phase motors still apply to single-phase motors. There are many exceptions that apply only to small-horsepower motors.

ACHIEVEMENT REVIEW

1. List the essential parts of a split-phase induction motor. _____

    _____

    _____

2. What happens if the centrifugal switch contacts fail to reclose when the motor stops?

    _____

    _____

3. Explain how the direction of rotation of a split-phase induction motor is reversed.

    _____

    _____

4. A split-phase induction motor has a dual-voltage rating of 115/230 volts. The motor has two running windings, each of which is rated at 115 volts, and one starting winding rated at 115 volts. Draw a schematic diagram of this split-phase induction motor connected for 230-volt operation.

5. Draw a schematic connection diagram of the split-phase induction motor in question 4 connected for 115-volt operation.

6. A split-phase induction motor has a dual-voltage rating of 115/230 volts. The motor has two running windings, each of which is rated at 115 volts. In addition, there are two starting windings and each of these windings is rated at 115 volts. Draw a schematic connection diagram of this split-phase induction motor connected for 230-volt operation.

7. What is the primary difference between a split-phase induction motor and a capacitor start, induction run motor? _____
    _____
    _____
    _____

8. If the centrifugal switch fails to open as a split-phase motor accelerates to its rated speed, what will happen to the starting winding? _____
    _____
    _____
    _____

9. What is one limitation of a capacitor start, induction run motor?
    _____
    _____
    _____

10. Insert the correct word or phrase to complete each of the following statements.
   a. A motor of one horsepower or less which is manually started and which is within sight of the starter location is considered to be protected by the _____.
   b. A motor of one horsepower or less which is manually started is considered within sight of the starter location if the distance is no greater than _____.
   c. The capacitor used with a capacitor start, induction run motor is used only to improve the _____.
   d. A capacitor start, induction run motor has a better starting torque than the

   _____.

REPULSION MOTORS

OBJECTIVES

After studying this unit, the student will be able to

- describe the basic steps in the operation of the following types of motors:
  repulsion motor
  repulsion start, induction run motor
  repulsion-induction motor
- state the basic construction differences among the motors listed in objective 1.
- compare the motors listed in objective 1 with regard to starting torque and speed performance.

Repulsion-type motors are divided into three distinct classifications: the repulsion motor; the repulsion start, induction run motor; and the repulsion-induction motor. Although these motors are similar in name, they differ in construction, operating characteristics, and industrial applications.

REPULSION MOTOR

A repulsion motor basically consists of the following parts:

Laminated stator core *with one winding*. This winding is similar to the main or running winding of a split-phase motor. The stator usually is wound with four, six, or eight poles.

Rotor consisting of *a slotted core into which a winding is placed.* The rotor is similar in construction to the armature of a dc motor. Thus, the rotor is called an armature. The coils which make up this armature winding are connected to a commutator. The commutator has segments or bars parallel to the armature shaft.

Carbon brushes contacting with the commutator surface. The brushes are held in place by a brush holder assembly mounted on one of the end shields. The brushes are connected together by heavy copper jumpers. The brush holder assembly may be moved so that the brushes can make contact with the commutator surface at different points to obtain the correct rotation and maximum torque output. There are two types of brush arrangements:

1. Brush riding – the brushes are in contact with the commutator surface at all times.
2. Brush lifting – the brushes lift at approximately 75 percent of the rotor speed.

Two cast steel end shields. These shields house the motor bearings and are secured to the motor frame.

Two bearings supporting the armature shaft. The bearings center the armature with respect to the stator core and windings. The bearings may be sleeve bearings or ball bearing units.

Cast steel frame into which the stator core is pressed.

Operation of a Repulsion Motor

The connection of the stator winding of a repulsion motor to a single-phase line causes a field to be developed by the current in the stator windings. This stator field induces a voltage and a resultant current in the rotor windings. If the brushes are placed in the proper position on the commutator segments, the current in the armature windings will set up proper magnetic poles in the armature.

These armature field poles have a set relationship to the stator field poles. That is, the magnetic poles developed in the armature are set off from the field poles of the stator winding by about 15 electrical degrees. Furthermore, since the instantaneous polarity of the rotor poles is the same as that of the adjacent stator poles, the repulsion torque created causes the rotation of the motor armature.

The three diagrams of figure 23–1 show the importance of the brushes being in the proper position to develop maximum torque. In figure 23–1A, no torque is developed when the brushes are placed at right angles to the stator poles. This is due to the fact that the equal induced voltages in the two halves of the armature winding oppose each other at the connection between the two sets of brushes. Since there is no current in the windings, flux is not developed by the armature windings.

In figure 23–1B, the brushes are in a position directly under the center of the stator poles. A heavy current exists in the armature windings with the brushes in this position, but there is still no torque. The heavy current in the armature windings sets up poles in the armature. However, these poles are centered with the stator poles and a torque is not created either in a clockwise or counterclockwise direction.

In figure 23–1C, the brushes have shifted from the center of the stator poles 15 electrical degrees in a counterclockwise direction. Thus, magnetic poles of like polarity are set up in the armature. These poles are 15 electrical degrees in a counterclockwise direction from the stator pole centers. A repulsion torque is created between the stator and the rotor field poles of like polarity. The torque causes the armature to rotate in a counterclockwise direction. A repulsion machine has a high starting torque, with a small starting current, and a rapidly decreasing speed with an increasing load.

The direction of rotation of a repulsion motor is reversed if the brushes are shifted electrical degrees from the stator field pole centers in a clockwise direction, figure 23–2. As a result, magnetic poles of like polarity are set up in the armature. These poles are 15

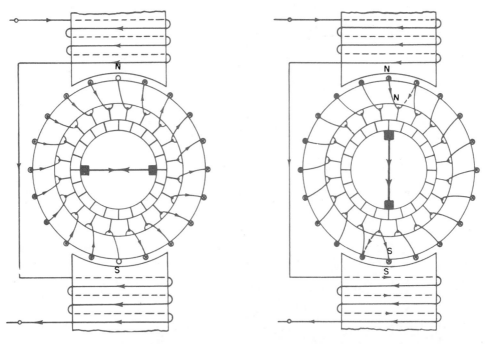

A. NO TORQUE CREATED, EQUAL VOLTAGE VALUES
   OPPOSE EACH OTHER (SOFT NEUTRAL)

B. NO TORQUE EVEN THOUGH CURRENT VALUE IN
   ARMATURE IS HIGH (HARD NEUTRAL)

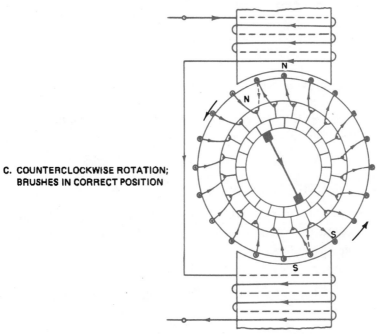

C. COUNTERCLOCKWISE ROTATION;
   BRUSHES IN CORRECT POSITION

Fig. 23–1   Repulsion motor operation

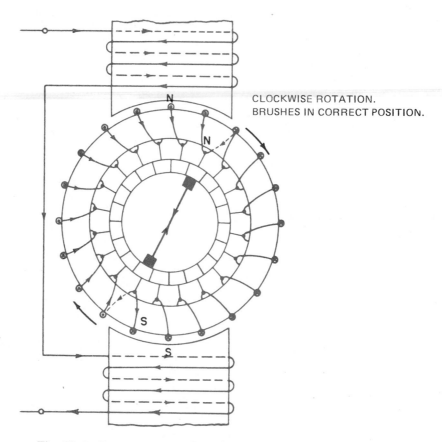

CLOCKWISE ROTATION.
BRUSHES IN CORRECT POSITION.

Fig. 23–2   Reversing the direction of rotation of a repulsion motor

electrical degrees in a clockwise direction from the stator pole centers. Repulsion motors are used principally for constant-torque applications, such as printing-press drives, fans, and blowers.

REPULSION START, INDUCTION RUN MOTOR

A second type of repulsion motor is the repulsion start, induction run motor. In this type of motor, the brushes contact the commutator at all times. The commutator of this motor is the more conventional axial form.

A repulsion start, induction run motor consists basically of the following parts.

Laminated stator core. This core has one winding which is similar to the main or running winding of a split-phase motor.

Rotor consisting of a slotted core into which a winding is placed. The coils which make up the winding are connected to a commutator. The rotor core and winding are similar to the armature of a dc motor. Thus, the rotor is called an armature.

Centrifugal device.

a. In the brush-lifting type of motor, there is a centrifugal device which lifts the brushes from the commutator surface at 75 percent of the rated speed. This device consists of governor weights, a short-circuiting necklace, a spring barrel, spring, push rods, brush holders, and brushes. Although high in first cost, this device does save wear and tear on brushes, and runs quietly. Figure 23–3 is an exploded view of the armature, radial commutator, and centrifugal device of the brush-lifting type of repulsion start, induction run motor.

b. The brush-riding type of motor also contains a centrifugal device which operates at 75 percent of the rated speed. This device consists of governor weights, a short-circuiting, necklace, and a spring barrel. The commutator segments are short circuited by this device, but the brushes and brush holders are not lifted from the commutator surface.

Commutator. The brush-lifting type of motor has a radial-type commutator (figure 23–3). The brush-riding type of motor has an axial commutator (figure 23–4).

Brush holder assembly.

a. The brush holder assembly for the brush-lifting type of motor is arranged so that the centrifugal device can lift the brush holders and brushes clear of the commutator surface.

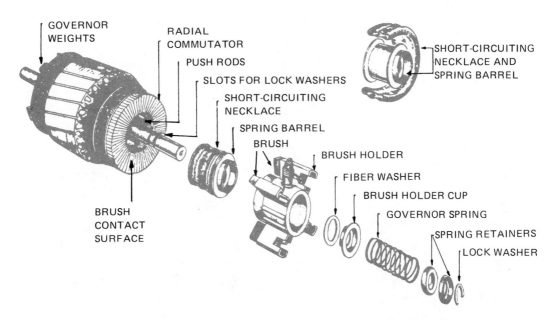

Fig. 23–3 An exploded view of a radial commutator and centrifugal brush-lifting device for a repulsion start, induction run motor

b. The brush holder assembly for the brush-riding type of motor is the same as that of a repulsion motor.

End shields, bearings, and motor frame. The parts are the same as those of a repulsion motor.

Operation of the Centrifugal Mechanism

Refer to figure 23–4 to identify the components of the centrifugal mechanism. The operation of this device consists of the following steps. As the push rods of the centrifugal device move forward, they push the spring barrel forward. This allows the short-circuiting necklace to make contact with the radial commutator bars thus are all short circuited. At the same time, the brush holders and brushes are moved from the commutator surface. As a result, there is no unnecessary wear on the brushes and the commutator surface and there are no objectionable noises caused by the brushes riding on the radial commutator surface.

The short-circuiting action of the governor mechanism and the commutator segments converts the armature to a form of squirrel-cage rotor and the motor operates as a single-phase induction motor. In other words, the motor starts as a repulsion motor and runs as an induction motor.

In the brush-riding type of motor, all axial commutator is used. The centrifugal mechanism (figure 23–4) consists of a number of copper segments which are held in place by a spring. This device is placed next to the commutator. When the rotor reaches 75 per-

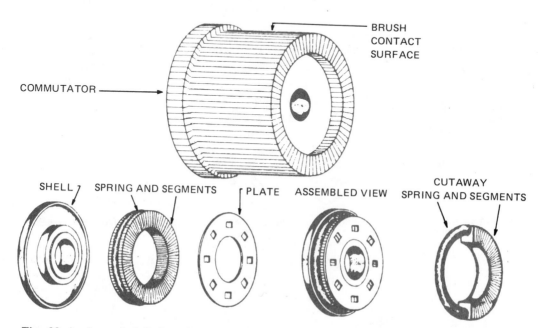

Fig. 23–4   An exploded view of a short-circuiting device for a brush-riding, repulsion start, induction run motor

cent of the rated speed, the centrifugal device short circuits the commutator segments. The motor then will continue to operate as an induction motor.

Operation of a Repulsion Start, Induction Run Motor

The starting torque is good for either the brush-lifting type or the brush-riding type of repulsion start, induction run motor. Furthermore, the speed performance of both types of motors is very good since they operate as single-phase induction motors.

Because of the excellent starting and running characteristics for both types of repulsion start, induction run motors, they were used for a variety of industrial applications, including commercial refrigerators, compressors, and pumps.

The direction of rotation for a repulsion start, induction run motor is changed in the same manner as that for a repulsion motor, that is, by shifting the brushes past the stator pole center 15 electrical degrees.

The symbol in figure 23–5 represents both a repulsion start, induction run motor and a repulsion motor.

Many repulsion start, induction run motors are designed to operate on 115 volts or 230 volts. These dual-voltage motors contain two stator windings. For 115-volt operation, the stator windings are connected in parallel; for 230-volt operation, the stator windings are connected in series. The diagram in figure 23–6 represent a dual-voltage, repulsion start, induction run motor. The connection table shows how the leads of the motor are connected for either 115-volt operation or 230-volt operation. These connections also can be used for dual-voltage repulsion motors.

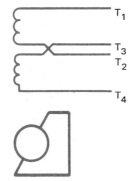

	L_1	L_2	TIE TOGETHER
LOW VOLTAGE	$T_1 T_3$	$T_2 T_4$	————
HIGH VOLTAGE	T_1	T_4	$T_2 T_3$

Fig. 23–5 Schematic diagram symbol of a repulsion start, induction run motor and a repulsion motor

Fig. 23–6 Schematic diagram of a dual-voltage, repulsion start, induction run motor

REPULSION-INDUCTION MOTOR

The operating characteristics of a repulsion-induction motor are similar to those of the repulsion start, induction run motor. However, the repulsion-induction motor has no centrifugal mechanism. It has the same type of armature and commutator as the repulsion motor, but it has a squirrel-cage winding beneath the slots of the armature.

Figure 23–7 shows a repulsion-induction motor armature with a squirrel-cage winding. One advantage of this type of motor is that it has no centrifugal device requiring maintenance. The repulsion-induction motor has a very good starting torque since it starts as a repulsion motor. At start up, the repulsion winding predominates; but, as the motor speed increases, the squirrel-cage winding is used most. The transition from repulsion to induction operation is smooth since no switching device is used. In addition, the repulsion-induction motor has a fairly constant speed regulation from no load to full load because of the squirrel-cage winding. The torque-speed performance of a repulsion-induction motor is similar to that of a dc compound motor.

A repulsion-induction motor can be operated on either 115 volts or 230 volts. The stator winding has two sections which are connected in parallel for 115-volt operation, and in series for 230-volt operation. The markings of the motor terminals and the connection arrangement of the leads is the same as in a repulsion start, induction run motor.

The symbol in figure 23–5 also represents a repulsion-induction motor (as well as a repulsion start, induction run motor and a repulsion motor.)

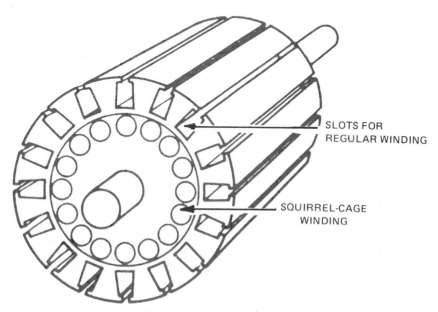

SLOTS FOR
REGULAR WINDING

SQUIRREL-CAGE
WINDING

Fig. 23–7   An armature of a repulsion-induction motor

NATIONAL ELECTRICAL CODE REGULATIONS

Regulations for the motor branch circuit overcurrent protection, motor running overcurrent protection, and wire sizes for motor circuits are given in *Article 430* of the National Electrical Code. Refer also to *Example 8, Chapter 9* of the Code.

SUMMARY

Repulsion motors are available in three basic designs: (1) repulsion motors, (2) repulsion start, induction run motors, and (3) repulsion-induction motors. Theses motors are easy to recognize because they are ac induction motors but use a commutator and brushes. The important points to remember is that the motors have neutral positions of the brush mountings that yield no motor movement. These neutral positions are referred to as *hard* or *soft neutral*. The brushes are shifted off neutral to give the motor the desired direction of rotation.

ACHIEVEMENT REVIEW

1. What is a repulsion motor, and how is rotation produced? _____

_____

_____

_____

_____

_____

2. Name one application of a repulsion motor. _____

_____

_____

3. Describe the operation of a repulsion start, induction run motor. _____

_____

_____

_____

_____

4. Explain the difference between the brush-lifting type of repulsion start, induction run motor and the brush-riding type of repulsion start, induction run motor.

_____

_____

_____

_____

5. A 2-hp, 230-volt, 12-ampere, single-phase repulsion start, induction run motor is connected directly across the rated line voltage.
    a. Determine the overcurrent protection for the branch circuit feeding this motor.

    _____

    _____

    b. Determine the running overcurrent protection for this motor.

    _____

    _____

6. What size wire is used for the branch circuit feeding the motor given in question 5?

    _____

    _____

7. Describe the construction of a repulsion-induction motor.

    _____

    _____

    _____

8. What is one advantage to the use of the repulsion-induction motor as compared to to the repulsion start, induction run motor?_____

    _____

    _____

9. Explain how the direction of rotation is changed for any one of the three types of single-phase repulsion motors covered in this unit. _____

    _____

    _____

    _____

10. Insert the correct word or phrase to complete each of the following statements.
    a. A repulsion-induction motor has a good _____ and a fairly good

    _____.

    b. A repulsion motor has a high starting torque and its speed rapidly decreases with

    _____.

    c. The centrifugal short-circuiting device on a repulsion start, induction run motor operates at approximately _____ of the rated speed.
    d. Both the repulsion start, induction run motor, and the repulsion-induction motor operate as _____ after they have accelerated to rated speed.

U•N•I•T
24
ALTERNATING-CURRENT SERIES MOTORS

OBJECTIVES

After studying this unit, the student will be able to

- describe the basic operation of a universal motor.

- explain how a single-field compensated universal motor operates.

- explain how a two-field compensated universal motor operates.

- describe two ways in which universal motors are compensated for excessive armature reaction under load.

- state the reasons why dc motors fail to operate satisfactorily from an ac source.

The electrician may consider a typical dc series motor or a dc shunt motor for operation on ac power supplies. It appears that such operation is possible since reversing the line terminals to a dc motor reverses the current and magnetic flux in both the field and armature circuits. As a result, the net torque of the motor operating from an ac source is in the same direction.

However, the operation of a dc shunt motor from an ac source is impractical because the high inductance of the shunt field causes the field current and the field flux to lag the line voltage by almost 90 degrees. The resulting torque is very low.

A dc series motor also fails to operate satisfactorily from an ac source because of the excessive heat developed by eddy currents in the field poles. In addition, there is an excessive voltage drop across the series field windings due to high reactance.

To reduce the eddy currents, the field poles can be laminated. To reduce the voltage loss across the field poles to a minimum, a small number of field turns can be used on a low reluctance core operated at low flux density. A motor with these revisions operates on either ac or dc and is known as a universal motor. Universal motors in small fractional horsepower sizes are used in household appliances and portable power tools.

CONCENTRATED FIELD UNIVERSAL MOTOR

A concentrated field universal motor usually is a salient-pole motor with two poles and a winding of relatively few turns. The poles and winding are connected to give opposite magnetic polarity. A field yoke of this type of motor is shown in figure 24–1.

DISTRIBUTED FIELD UNIVERSAL MOTORS

The two types of distributed field universal motors are: the single-field compensated motor and the two-field compensated motor.

The field windings of a two-pole, single-field compensated motor resemble the stator winding of a two-pole, split-phase ac motor. A two-field compensated motor has a stator containing a main winding and a compensating winding spaced 90 electrical degrees apart. The compensating winding reduces the reactance voltage developed in the armature by the alternating flux when the motor operates from an ac source. Figure 24-2 is the schematic diagram of a compensated universal motor.

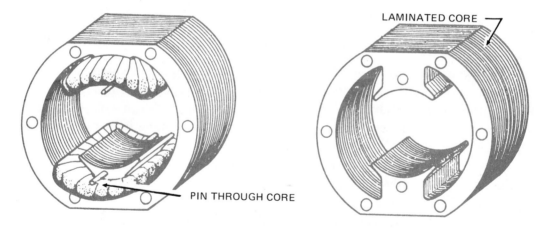

Fig. 24–1   Field core of a two-pole universal motor

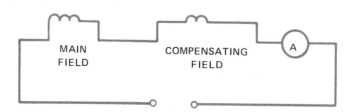

Fig. 24–2   Schematic diagram of a compensated universal motor

THE ARMATURE

The armature of a typical universal motor resembles the armature of a typical dc motor except that a universal motor armature is slightly larger for the same horsepower output.

CONSTRUCTION FEATURES OF UNIVERSAL MOTORS

The frames of universal motors are made of aluminum, cast iron, and rolled steel. The field poles are generally bolted to the frame. Field cores consist of laminations pressed together and held by bolts. The armature core is also laminated and has a typical commutator and brushes. End plates resemble those of other motors except that in many universal motors one end plate is cast as part of the frame. Both ball and sleeve bearings are used in universal motors.

SPEED CONTROL

Universal motors operate at approximately the same speed on dc or single-phase ac. Since these motors are series wound, they will operate at excessive speed at a no-load condition. As a result, they usually are permanently connected to the devices being driven. Universal motors are speed regulated by inserting resistance in series with the motor. The resistance may be tapped resistors, rheostats, or tapped nichrome wire coils wound over a single field pole. In addition, speed may be controlled by varying the inductance through taps on one of the field poles. Gear boxes are also used.

Speed control of series motors can also be accomplished by using electronic speed controls. The concept is the same as used series voltage drops; that is, the voltage to the motor is reduced to give a reduced speed. This can be done by using SCRs or triacs to alter the voltage available to the motor.

DIRECTION OF ROTATION

The direction of rotation of any series wound motor can be reversed by changing the direction of the current in either the field or the armature circuit. Universal motors are sensitive to brush position and severe arcing at the brushes will result from changing the direction of rotation without shifting the brushes to the neutral (sparkless) plane, or adding a compensating winding.

CONDUCTIVE COMPENSATION

Ac motors rated at more than 1/2 horsepower are used to drive loads requiring a high starting torque. Two methods are used to compensate for excessive armature reaction under load. In the conductively compensated type of motor, an additional compensating winding is placed in slots cut directly into the pole faces. The strength of this field increases with an increase in load current and thus minimizes the distortion of the main field flux by the armature flux (called armature reaction-discussed in DC Motor Unit-).

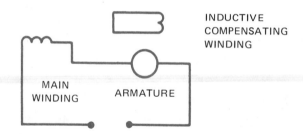

Fig. 24–3 Connections for an inductively compensated universal motor

The compensating winding is connected in series with the series field winding and the armature, as shown in figure 24-3. Although conductively compensated motors have a high starting torque, the speed regulation is poor. A wide range of speed control is possible with the use of resistor-type starter-controllers.

INDUCTIVE COMPENSATION

Armature reaction in ac series motors also may be compensated with an inductively coupled winding which acts as a short-circuited secondary winding of a transformer. This winding is placed so that it links the cross-magnetizing flux of the armature which acts as the primary winding of a transformer, Figure 24-3 is the schematic diagram of an inductively compensated universal motor. Since the magnetomotive force of the secondary is nearly opposite in phase, and equal in magnitude, to the primary magnetomotive force, the compensating winding flux nearly neutralizes the armature cross flux. This type of motor cannot be used on dc current. Because of its dependency on induction, the operating characteristics of an inductively compensated motor are very similar to those of the conductively compensated motor.

SUMMARY

AC series motors are conduction motors, just as the series dc motors. The construction is slightly different since the magnetic field changes affect the inductance of the iron. The principle of operation is the same as that of the series dc motor. The armature keeps the same magnetic polarity of the rotor, reacting with the same magnetic field of the stator through the process of commutation.

ACHIEVEMENT REVIEW

A. Completely answer the following questions.

1. a. Describe the basic differences in construction between the concentrated field and the distributed field types of universal motors, b. Draw the schematic diagram for each type of motor.

a. _____

_____

_____

_____

b.

2. What is the function of the compensating winding in a two-field compensated universal motor? _____

_____

_____

3. Describe three methods of controlling the speed of universal motors. _____

_____

_____

_____

4. Why does a universal motor spark excessively at the commutator if its direction of rotation is -reversed? _____

_____

_____

5. A dc series motor operates unsatisfactorily on ac. What are the primary reasons for this fact? _____

_____

_____

B. Select the correct answer for each of the following statements and place the corresponding letter in the space provided.

6. The operation of a dc shunt motor from an ac source is impractical because _____
    a. too much torque is developed at startup.
    b. the starting current is too high.
    c. the shunt field inductance is too high.
    d. the shunt field inductance is too low.

7. A series dc motor fails to operate satisfactorily on ac due to _____
    a. eddy currents and high field voltage drop.
    b. excessive heat and low field voltage drop.
    c. low reactance of the armature and field.
    d. high armature reluctance and low field reactance.

8. The frames of universal motors are made of _____
    a. rolled steel.   c. aluminum.
    b. cast iron.   d. all of these.

9. A compensating winding
    a. increases the reactance in the armature on ac.
    b. reduces the reactance in the armature on ac.
    c. reduces the reactance in the armature on dc.
    d. increases the reactance in the armature on dc.

10. After changing the direction of rotation of a universal motor, the _____
    a. brushes must be rotated for sparkless commutation.
    b. field connections must be shifted.
    c. field reactance must be decreased.
    d. field reactance must be increased.

25

SELSYN UNITS

OBJECTIVES

After studying this unit, the student will be able to

- describe the operation of a simple selsyn system.
- describe the operation of a differential selsyn system.
- list several advantages of selsyn units.

The word *selsyn* is an abbreviation of the words self-synchronous. Selsyn units are special ac motors used primarily in applications requiring remote control. Small selsyn units transmit meter readings or values of various types of electrical and physical quantities to distant points. For example, the captain on the bridge of a ship may adjust the course and speed of the ship; at the same moment, the course and speed changes are transmitted to the engine room by selsyn units. On the engine telegraph system, mechanical positioning of a control transmits electrical angular information to a receiving unit. Similarly, readings of mechanical and electrical conditions in other parts of the ship can be recorded on the bridge by selsyn units. These units are also referred to as *synchros,* and are known by various trade names.

STANDARD SELSYN SYSTEM

A selsyn system consists of two three-phase induction motors. The normally stationary rotors of these induction motors are interconnected so that a manual shift in the rotor position of one machine is accompanied by an electrical rotor shift in the other machine in the same direction and of the same angular displacement as the first unit.

Figure 25–1 shows a simple selsyn system for which the units at the transmitter and receiver are identical. The rotors of these units are two pole and must be excited from the same ac source. The three-phase stator windings are connected to each other by three leads between the transmitter and the receiver units. The rotor of each machine is called the *primary* and the three-phase stator winding of each machine is called the *secondary*. A rotor for a typical selsyn unit is shown in figure 25–2.

When the primary excitation circuit is closed, an ac voltage is impressed on the transmitter and receiver primaries. If both rotors are in the same position with respect to their stators, no movement occurs. If the rotors are not in the same relative position, the freely movable receiver rotor will turn to assume the same position as the transmitter rotor.

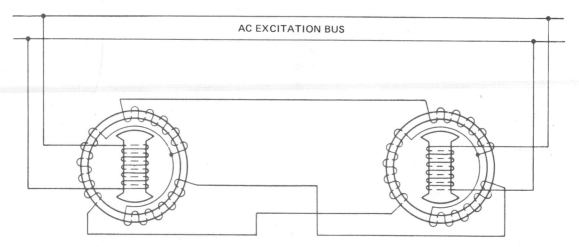

Fig. 25–1   **Diagram of selsyn motors showing interconnected stator and rotor windings connected to excitation source**

Fig. 25–2   **Wound rotor with oscillation damper and slip rings for selsyn units (*Photo courtesy of General Electric Company*)**

If the transmitter rotor is turned, either manually or mechanically, the receiver rotor will follow at the same speed and in the same direction.

The self-synchronous alignment of the rotors is the result of voltages induced in the secondary windings. Both rotors induce voltages into the three windings of their stators. These voltages vary with the position of the rotors. If the two rotors are in the same relative position, the voltages induced in the transmitter and receiver secondaries will be equal and opposite. In this condition, current will not exist in any part of the secondary circuit.

If the transmitter rotor is moved to another position, the induced voltages of the secondaries are no longer equal and opposite and currents are present in the windings. These currents establish a torque which tends to return the rotors to a synchronous position. Since the receiver rotor is free to move, it makes the adjustment. Any movement of the transmitter rotor is accompanied immediately by an identical movement of the receiver rotor.

DIFFERENTIAL SELSYN SYSTEM

Figure 25–3 is a diagram of the connections of a differential selsyn system consisting of a transmitter, a receiver, and a differential unit. This system produces an angular indication of the receiver. The indication is either the sum or difference of the angles existing at the transmitter and differential selsyns. If two selsyn generators, connected through a differential selsyn, are moved manually to different angles, the differential selsyn will indicate the sum or difference of their angles.

A differential selsyn has a primary winding with three terminals. Otherwise, it closely resembles a standard selsyn unit. The three primary leads of the differential selsyn are brought out to collector rings. The unit has the appearance of a miniature wound-rotor, three-phase induction motor. The unit, however, normally operates as a single-phase transformer.

The voltage distribution in the primary winding of the differential selsyn is the same as that in the secondary winding of the selsyn exciter. If any one of the units is fixed in position and a second unit is displaced by a given angle, then the third unit which is free to rotate will turn through the same angle. The direction of rotation can be reversed by interchanging any pair of leads on either the rotor or stator winding of the differential selsyn.

If any two of the selsyns are rotated simultaneously, the third selsyn will turn through an angle equal to the algebraic sum of the movements of the two selsyns. The algebraic sign of this value depends on the direction of rotation of the rotors of the two selsyns, as well as the phase rotation of their windings.

The excitation current of the differential selsyn is supplied through connections to one or both of the standard selsyns to which the differential selsyn is connected. In general, the excitation current is supplied to the primary winding only. In this case, the selsyn connected to the differential stator supplies this current and must be able to carry the extra load without overheating. A particular type of selsyn, known as an *exciter selsyn,* is used to supply the current. The exciter selsyn can function in the system either as a transmitter or a receiver.

ADVANTAGES OF SELSYN UNITS

Selsyn units are compact and rugged and provide accurate and very reliable readings. Because of the comparatively high torque of the selsyn unit, the indicating pointer does not oscillate as it swings into position. Internal mechanical dampers are used in selsyn receivers to prevent oscillation during the synchronizing procedure and to reduce any tendency of the receiver to operate as a rotor. The operation of the receiver is smooth and continuous and is in agreement with the transmitter. In addition, the response of the receiver to changes in position at the transmitter is very rapid.

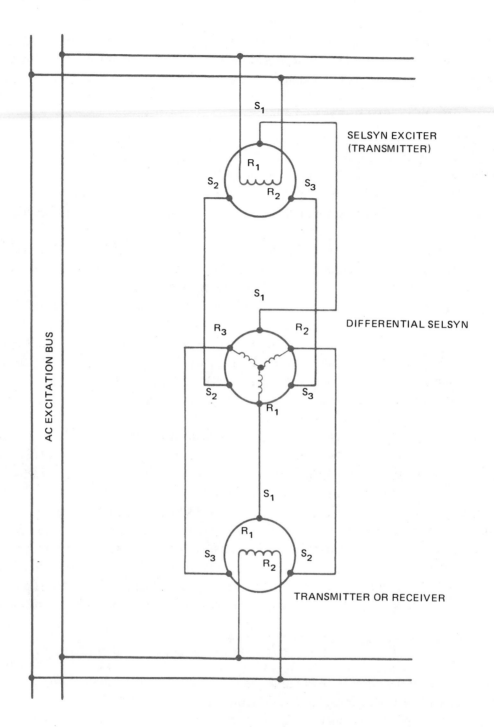

Fig. 25–3   A schematic diagram of differential selsyn connections

In the event of a power failure, the indicator of the receiver resets automatically with the transmitter when power is received. Calibration and time-consuming checks are unnecessary.

A number of advantages are offered by selsyn systems.

- The indicators are small and compact and can be located where needed.
- The simple installation requires running a few wires and bolting the selsyn units in place.
- Selsyn units can be used to indicate either angular or linear movement.
- Selsyn units control the motion of a device at a distant point by controlling its actuating mechanism.
- One transmitter may be used to operate several receivers simultaneously at several distant points.

SUMMARY

The selsyn system is also referred to as a syncro system. The self-synchronous system allows one rotor to act as a transmitter and another rotor to act as a receiver to follow the transmitter. There are several variations to allow the receiver to follow at some angle determined by a differential selsyn.

ACHIEVEMENT REVIEW

Select the correct answer for each of the following statements and place the corresponding letter in the space provided.

1. Selsyn transmitters and receivers resemble   _____
   a. repulsion-induction motors.
   b. three-phase, two-pole induction motors.
   c. three-phase, four-pole induction motors.
   d. synchronous machines.

2. When the primary excitation circuit is closed, ac voltage is
   impressed on the   _____
   a. transmitter and receiver primaries.
   b. transmitter rotor and the transmitter stator windings.
   c. transmitter rotor and the receiver stator windings.
   d. stator windings of both instruments.

3. A differential selsyn unit differs from a selsyn transmitter or
   receiver in that it requires   _____
   a. three-phase power for excitation.
   b. an ac line connection to the stator winding.
   c. dc on the rotor winding.
   d. three connections to the rotor winding.

4. If the rotors of the two selsyn units in a selsyn indicating system are in exactly corresponding positions, the current in the secondary winding is _____
   a. within quadrature with the primary current.
   b. in phase with the primary current.
   c. zero.
   d. less than the normal primary current.

5. Selsyn units are also referred to as _____
   a. synchros.
   b. induction motors.
   c. wound-rotor motors.
   d. all of these.

6. The stator of the transmitter is directly connected to the stator of the receiver unit when a differential is not used. _____
   a. true
   b. false

7. In the transmitter and receiver system, the ac excitation is applied to the

   _____
   a. stator winding.
   b. stator and the rotor windings.
   c. rotor winding only
   d. none of these.

8. Cite several advantages of a selsyn system. _____

   _____

   _____

   _____

   _____

   _____

26

MOTOR MAINTENANCE

OBJECTIVES

After studying this unit, the student will be able to

- perform routine inspection and maintenance checks of three-phase motors.
- perform the following simple tests:

  measure insulation resistance

  use a growler to locate short-circuited coils

  continuity checks for open-circuited coil

  balance test to determine phase currents under load

  speed variation
- replace and lubricate sleeve and ball bearings, according to manufacturers' directions.
- lubricate motors, according to manufacturers' directions.

PREVENTIVE MAINTENANCE

Most electrical equipment requires planned inspection and maintenance to keep it in proper working condition. Periodic inspections prevent serious damage to machinery by locating potential trouble areas. Observant personnel will make full use of their senses to diagnose and locate problems in electrical machinery: the sense of *smell* directs attention to burning insulation; the sense of *touch* detects excessive heating in windings or bearings; the sense of *hearing* detects excessive speed or vibration; the sense of *sight* detects excessive sparking and many mechanical faults.

Sensory impressions usually must be supplemented by various testing procedures to localize the trouble. A thorough understanding of electrical principles and the efficient use of test equipment is important to the electrician in this phase of troubleshooting.

PERIODIC INSPECTIONS

The ideal motor maintenance program aims at preventing breakdowns rather than repairing them. Systematic and periodic inspections of motors are necessary to ensure best operating results. In a good preventive maintenance program with detailed checks, the person in charge should have a record card on file for every motor in the plant. Entries on the card should include inspection dates, descriptions of repairs, and the costs involved. When the record indicates that a motor has undergone excessive and/or costly repairs, the causes can be determined and corrected.

Inspection records also serve as a guide to indicate when motors should be replaced because of their high cost of operation. They also reveal faulty operating conditions, such as misapplication or poor drive engineering.

Inspection and servicing should be systematic. However, the frequency of inspections and the degree of thoroughness may vary, as determined by the plant maintenance engineer. Such determinations are based on 1) the importance of the motor in the production scheme (if the motor fails, will production be slowed seriously, or stopped?), 2) the percentage of the day the motor operates, 3) the nature of the service, and 4) the motor's environment. An inspection schedule, therefore, must be flexible, and adapted to the needs of each plant. Equipment manufacturers' specifications and procedures should be consulted and followed.

The following schedule, which covers both ac and dc motors, is based on average conditions insofar as operational use and cleanliness are concerned. (Where dust and dirty conditions are extremely severe, open motors may require a certain amount of cleaning every day.)

EVERY WEEK

1. Examine commutator and brushes, ac and dc.
2. Check oil level in bearings.
3. See that oil rings turn with shaft.
4. See that exposed shaft is free of oil and grease from bearings.
5. Examine the starter switch, fuses, and other controls; tighten loose connections.
6. See that the motor is brought up to speed in normal time.

EVERY SIX MONTHS

1. Clean motor thoroughly, blowing out dirt from windings, and wipe commutator and brushes.
2. Inspect commutator clamping ring.
3. Check brushes and replace any that are more than half worn.
4. Examine brush holders, and clean them if dirty. Make certain that brushes ride free in the holders.
5. Check brush pressure.
6. Check brush position.
7. Drain, wash out, and replace oil in sleeve bearings.

8. Check grease in ball or roller bearings.

9. Check operating speed or speeds.

10. See that end play of shaft is normal.

11. Inspect and tighten connections on motor and control.

12. Check current input and compare it with normal.

13. Examine drive, critically, for smooth running, absence of vibration, and worn gears, chains, or belts.

14. Check motor foot bolts, end-shield bolts, pulley, coupling, gear and journal set-screws, and keys.

15. See that all covers, and belt and gear guards are in place, in good order, and securely fastened.

ONCE A YEAR

1. Clean out and renew grease in ball or roller bearing housings.

2. Test insulation by megohmmeter.

3. Check air gap.

4. Clean out magnetic dirt that may be clinging to poles.

5. Check clearance between shaft and journal boxes of sleeve bearing motors to prevent operation with worn bearings.

6. Clean out undercut slots in commutator. Check the commutator for smoothness.

7. Examine connections of commutator and armature coils.

8. Inspect anrmature bands.

MEASUREMENT OF INSULATION RESISTANCE

The condition of insulation is an important factor in the maintenance of motors and alternators. Moisture, dirt, chemical fumes, and iron particles all cause deterioration of the insulation used on the windings of stators and rotors. Motors operating under adverse conditions require periodic tests to insure continuous and satisfactory operation. Although severe conditions can be detected by touch, sight, or smell, often it is necessary to use more accurate measures of the condition of insulation at any given time.

The value of insulation resistance in *megohms (1,000,000 ohms)* is used as an indication of insulation efficiency. Successive readings taken and recorded under the same test conditions will document the insulation history of a unit and will serve as an index of insulation deterioration. A *megohmmeter,* commonly called a Megger® (figure 26–1A and B) is used to measure insulation resistance. In general, the instrument develops a voltage which is applied to the insulation path. The amount of current in this path is shown on

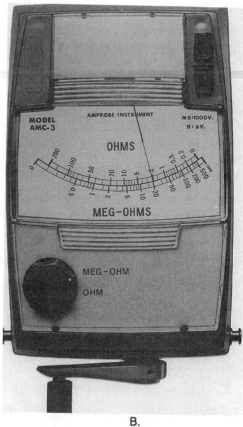

A.

B.

Fig. 26–1 A) Connections for testing motor winding resistance to the motor frame B) Megohmmeter used with hand crank operation (Photos from Keljik, *Electric Motors and Motor Controls*, copyright 1995 by Delmar Publishers)

a sensitive *microammeter* calibrated in megohms. When the instrument leads are not connected, the microammeter reading should be infinity. Specific operating instructions are provided with the instrument by the manufacturer.

The megohmmeter ground lead is connected to the frame of the machine. The ungrounded lead is connected to any metallic part of the winding being tested, such as any terminal of the coil circuit. No other external paths must exist in parallel circuits. For this reason, the windings being tested should be isolated by disconnecting them from other parts of the circuit. The megohmmeter reading then is the resistance of the insulation between the number of coils in the circuit and the frame of the machine.

Major motor manufacturers apply insulation resistance for 660 volts for low-voltage motors below 600 volts. A method of determining the normal insulation resistance value is given in the following formula. It is a nonexact formula derived from experience. It is believed that the Institute of Electrical and Electronic Engineers, Inc. (IEEE) accepted this general theory from a manufacturer of megohmmeters.

Caution: Disconnect electronic devices attached to units under test.

$$\text{Megohms} = \frac{\text{Rated Voltage of Machine}}{\dfrac{\text{Rating in kVA}}{100} + 1{,}000}$$

A very low value of insulation resistance indicates defective insulation. The electrician should begin an immediate check to localize the defective insulation. This is done by disconnecting various coils from a series, parallel, wye, or delta combination and repeating the insulation resistance test on each isolated coil. A very low value of resistance may indicate a grounded coil (complete breakdown of insulation at some point).

A value of insulation resistance slightly below the recommended approximate value does not necessarily indicate that immediate repair is required. The electrician should take a series of readings at weekly intervals to detect any progressive decrease or sudden drop in the insulation resistance. If the resistance continues to decrease, then the fault is to be located without delay. A slightly low but constant value of resistance should not cause concern. NOTE: If a motor has been reading 10 megohms on periodic tests and suddenly drops to 0.2 megohms, the motor may be unsafe to start. New motors read around infinity (the highest resistance on the meter scale), or slightly lower, but, during use, this resistance lowers to a more steady reading.

TESTING FOR SHORT-CIRCUITED COILS

Open-circuited coils on rotors or stators can be located by continuity (one end to the other) tests. Short-circuited coils are located easily by the use of a growler. A *growler* is an instrument consisting of an electromagnetic yoke and winding excited from an ac source. The yoke is placed across a section of slots containing the winding being tested. The yoke winding acts as the primary winding of a transformer and the winding being tested acts as the secondary (figure 26–2).

If a turn or coil is short circuited, the resulting current rise in the primary (yoke) circuit is indicated on the ammeter. If the current is permitted to exist for a short period of time, the defective turn or coil can be identified by the heat developed at the defective point. The stator windings of both alternators and motors can be tested by this method.

The field coils of alternators can be tested using the voltage drop method described for dc machinery testing. With a given value of current in the field circuit, the voltage drops on individual field coils should be approximately equal. If there is a voltage drop difference in excess of 5 percent, the coil should be investigated for shorted turns. The presence of full line voltage across a single coil indicates an open circuit in that coil.

The field coils of alternators can be checked for impedance by applying a high frequency low voltage to each coil and measuring the current. The currents in the coils should be equal. The presence of a high current usually means that there are shorted turns somewhere in the coil.

BALANCE TEST

The current in the individual phases of three-phase motors must be equal. Figure 26–3 shows the connections necessary to make a simple balance test which measures the phase currents under load. This test may be made in the electric shop before a motor is installed.

The fused three-pole switch at the left of the diagram is used to start the motor. The second three-pole switch removes the ammeter from the circuit during the starting period when the current input is very high. This switch is closed before the motor is started. When the motor reaches its rated speed, the ammeter switch is opened. The current in each phase, therefore, is indicated on the ammeter. For a motor operating normally, the

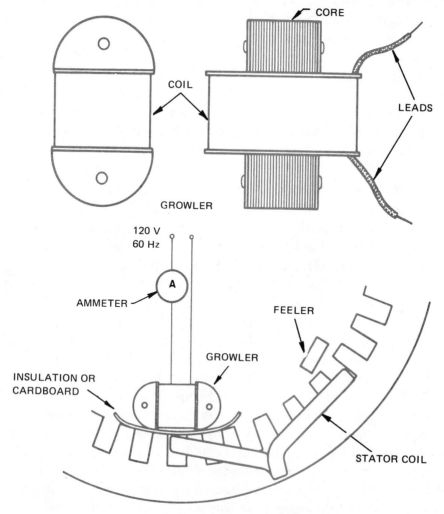

Fig. 26–2 A growler for testing shorted coils

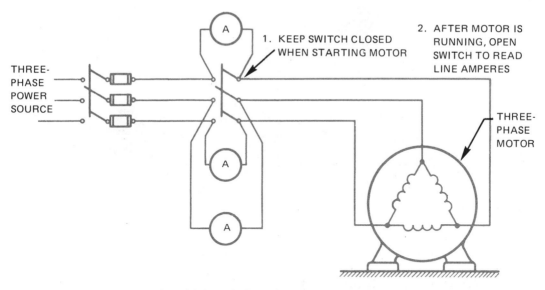

Fig. 26–3   A balanced current motor test

three line currents are equal. A high reading in one phase may indicate shorted turns. If one phase shows no current, then the motor is operating as a single-phase motor. High but equal current readings in all three phases indicate an overloaded motor.

Figure 26-4 shows the use of a more convenient method of making a balance test. A clamp-on ammeter is used to take readings in each phase of a motor in actual operation under normal load.

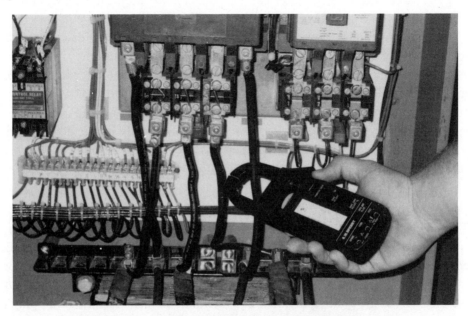

Fig. 26–4   Use of clamp-on ammeter to test motor power system

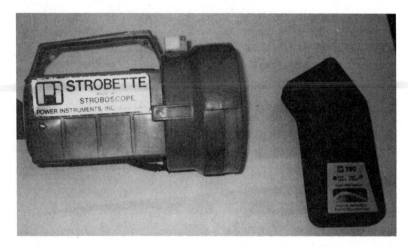

Fig. 26–5   Strobe tachometer and a photo reflective tachometer

MEASUREMENT OF SPEED VARIATIONS

Deviations from the rated speed of a motor under load are an indication of improper mechanical loading or faulty conditions within the motor. A *tachometer* (figures 26-5 and 26-6) is an instrument used to check the speed of a motor. Other types of instruments are also used. The tachometer in figure 26–6 is held by hand to the end of the motor shaft to measure shaft speed.

SQUIRREL CAGE ROTORS

The bars of a squirrel-cage rotor can be broken if the motor is subjected to severe jarring, vibration, or overheating. Bars that are dislodged must be put back in place securely so that movement is impossible. The presence of broken bars can be detected by the use of a growler. (The rotor must be removed to make this test.)

A test can be made which does not require the removal of the rotor to detect broken or open rotor bars. This test consists of exciting one phase of the stator winding with 25 percent of its normal voltage. Enough voltage is used to give a suitable indication on an ammeter in series with the winding. By turning the rotor slowly by hand, any variations in stator current can be observed on the ammeter. Any current variation in excess of 3 percent usually indicates open bars in the rotor (Figure 26-7).

BEARINGS

The type of bearings used in a motor depends on the cost of the motor and the characteristics of the load. Sleeve and ball bearings are used in both ac and dc motors. Excessive wear on the bearings reduces the concentricity of the stator and rotor sections. In small motors, with the power off, excessive wear can be detected by manually attempting to move the shaft of the rotating member in a lateral direction. The amount of play in the

Fig. 26–6   Mechanical tachometer used to measure shaft RPM

shaft is an indication of bearing wear. For large motors, bearing wear and the resulting deviation in concentricity of the stator and rotor can be detected by measuring the air gaps over several points around the periphery of the rotor. Severe bearing wear on both large and small motors may result in actual contact between the rotor and the stator.

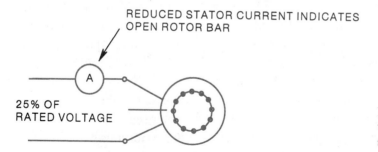

Fig. 26–7   An ammeter used to test for open rotor bars in a squirrel cage rotor

The motor must be disassembled to repair bearings. This type of repair requires special tools and the job should not be attempted without them.

A sleeve bearing is removed by dismantling the baffles inside the end shield, removing the oil well cover plates, and removing the oil ring clips. The bearing lining is then tapped out using a short length of pipe stock or a special split fitting which locks inside the oil ring slot.

Ball Bearings

Ball bearings are press fitted to the motor shaft and usually should be removed from the shaft only when it is necessary to replace the bearing.

To inspect the ball bearings, the end bells are removed and the rotor, the rotor shaft, and the bearing assembly are taken from the stator. In some motors, the bearing housings have removable bearing caps so that it is possible to remove the bearing without removing the end bells.

If a ball bearing must be replaced, a bearing puller usually is used to prevent damage to the shaft. The electrician must be very careful when placing a new bearing on the shaft so that neither the bearing nor the shaft is damaged (figure 26–8). A ball bearing race must be placed on the shaft so that the race is exactly square with the shaft. The shaft must be in perfect condition since even the slightest burr will cause trouble. **Pressure must not be applied to the outer race of the bearing**. Pressure applied to install the bearing must be applied evenly on the diameter of the inner race (figure 26–9). Light tapping is recommended. A piece of pipe stock slightly greater than the shaft diameter is used to press on the new bearing. The bearing can be warmed to a temperature of 150° F to simplify the process.

New ball bearings must not be cleaned prior to installation. Dust or dirt must not enter the bearing during the installation.

LUBRICATION

Several methods of lubricating motors are used. Small motors with sleeve bearings have oil holes with spring covers. These motors should be oiled periodically with a good grade of mineral oil. Oil with a viscosity of 200 seconds Saybolt (approximately SAE 10) is recommended.

The bearings of larger motors often are provided with an oil ring which fits loosely in a slot in the bearing. The oil ring picks up oil from a reservoir located directly under the ring. Under normal operating conditions, the oil should be replaced in the motor at least once a year. More frequent oil replacement may be necessary in motors operating under adverse conditions. In all cases, avoid excessive oiling; insufficient oil can ruin a bearing but excessive oil can cause deterioration of the insulation of a winding.

Many motors are lubricated with grease. Periodic replacement of the grease is recommended. In general, the grease should be replaced whenever a general overhaul is indicated, or sooner if the motor is operated under severe operating conditions.

WRONG:

BEARING SHOULD NOT
BE FORCED ON SHAFT
BY TAPPING ON THE
OUTER RINGS. IT
SHOULD NOT BE FORCED
ON A BADLY WORN
SHAFT OR ON A SHAFT
THAT IS TOO LARGE.

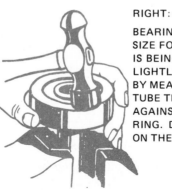

RIGHT:

BEARING IS PROPER
SIZE FOR SHAFT AND
IS BEING TAPPED
LIGHTLY INTO PLACE
BY MEANS OF A METAL
TUBE THAT FITS
AGAINST THE INNER
RING. DO NOT POUND
ON THE BEARING.

Fig. 26–8   Bearing installation

Fig. 26–9   Single-row, snap ring ball
bearing (*Photo courtesy of New Departure
Division, General Motors Corporation*)

Grease can be removed by using a light mineral oil heated to 165° F or a solvent. Any grease-removing solvents should be used in a well-ventilated work area.

Ball Bearing Lubrication

The following table indicates recommended intervals between the regreasing of ball bearing units. It is important that the correct amount and type of grease be used in ball bearings. Too much grease can cause overheating.

Horsepower			
Service	1/4-7 1/2	10-40	50-150
Easy	7 yrs.	5 yrs.	3 yrs.
Standard	5 yrs.	2 yrs.	1 yr.
Severe	3 yrs.	1 yr.	6 mos.
Very severe	6 mos.	3 mos.	3 mos.

Fractional horsepower motors often contain sealed bearings. With normal service, these bearings do not require regreasing. When regreasing is required for unshielded bearings, the manufacturers' specifications and directions must be followed.

Some motors with ball bearings are provided with pressure fittings and a grease gun is used to lubricate the bearings. Remove the bottom plug when doing this. Because of the wide variation in the design of industrial motors, the electrician should consult the comprehensive lubrication manuals published by electrical machinery manufacturers to insure the proper lubrication of all types of motors.

BALL AND ROLLER BEARING MAINTENANCE

Cleaning Out Old Grease

It is best to remove the bearing, if possible. However, when cleaning a bearing in place, remove as much of the old grease as possible, using a rag or brush free from dirt or dust. Flush out the housing with clean, hot kerosene (110° F to 125° F); clean, new oil; or solvent.

After the grease has been removed, flush out the bearing with light mineral oil to prevent rust and to remove all traces of cleaning fluid. Allow to drain thoroughly before adding new grease.

When the bearing is removed from the case, wash out the hardened, rancid grease from the housing and bearing as follows:

1. Put on safety glasses, or a face shield, for protection.

2. Soak the bearing in hot kerosene, then remove it.

3. Rotate the bearing slowly by air hose. (Fast rotation may score the balls without lubrication.)

4. Dip the bearing in clean kerosene, light oil, or solvent.

5. Rotate the bearing slowly again by air hose.

6. Wash the bearing again with clean kerosene or solvent.

7. Rotate the bearing in hand, and check for smoothness.

8. If the bearing is smooth, repack it with grease.

9. If the bearing is not smooth, but is in good condition, there is still some hardened grease in it. Repeat operations 6 and 7.

Lubricating Bearings with Pressure Relief Plug

1. Wipe fitting and plug clean.

2. Remove relief plug in bottom of bearing (figure 26–10).

3. With shaft in motion (if possible), force grease into the top grease fitting, catching old grease in a pan. Add grease until new grease appears at the pan.

4. With a screwdriver, open the relief hole. Do *not* push the screwdriver into the gearing housing.

5. Allow bearing to run with relief plug out to remove pressure.

6. After the grease stops running out, and there is no longer any pressure in the bearing, replace the relief plug.

Lubricating Bearings without Relief Plug

1. Install a grease fitting with safety vents.

2. With the motor running, pump in grease slowly until a slight bleed shows around the seal or safety vent.

3. If necessary to lubricate while bearing is standing, fill the bearing with grease from one-fourth to one-half of its capacity.

    NOTE: It is important not to overgrease, and grease must be kept clean.

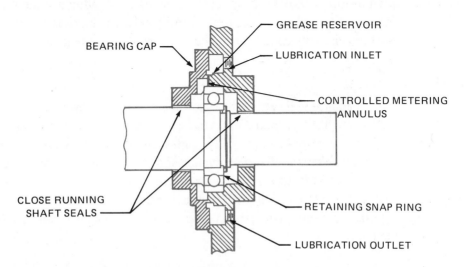

Fig. 26–10   Ball bearing housing assembly. Note the lubrication inlet and outlet (*Courtesy of General Electric Company*)

Relubrication with Oil

Oil is always subject to gradual deterioration from use, and contamination from dirt and moisture. Because of this, regular intervals for cleaning bearings must be maintained.

1. After draining the used oil, flush out the bearing. This can be done by using a new charge of the lubricant used for regular lubrication. Run the machine for three to five minutes, and drain again. Units used where there is sawdust or dirty conditions may require two flushings.

2. Fill with new oil to the proper level. Be sure new oil is from a clean container, and that no dirt is pushed into the filler plug while filling. Lubricant must be kept clean.

3. Check the seals to see that they are effective to prevent leakage or the entry of outside dirt to the bearing. It is important to keep the lubricant clean and the bearing flushed out so that it will be clean. Always read and follow the manufacturer's instructions whenever possible.

OILS FOR GEAR MOTORS

A gear motor is a self-contained drive made up of a ball bearing motor and a speed reducing gear box. It is designed to take advantage of the electrical efficiency of the high speed motor and the transmission efficiency of gears.

The front motor bearings are generally grease lubricated and require the same attention as standard ball bearing motors. The rear bearings, gear box bearings, and the gears themselves are almost always splash lubricated from the same oil supply reservoir in the lower section of the gear unit.

Oil seals at each bearing prevent oil leakage into the motor windings and out along the drive shaft. The precision cut gears require carefully selected lubricating oils. Use only top grade oils of the viscosity called for by the manufacturer of the gear motor.

SUMMARY

Proper motor maintenance will extend the life of the motor. Preventive maintenance will help you spot potential problems and correct them before the motor breaks down and causes delays in productions or failures at critical times. If a motor does fail, it may be possible to repair minor problems in the maintenance shop and to recondition the motor for use as a spare. Follow manufacturers directions when lubricating the motor bearings to keep the motor running smoothly without excessive current draw or excessive heat. Check for proper ventilation, proper voltage and proper current when performing maintenance checks on the motor.

ACHIEVEMENT REVIEW

A. Select the correct answer for each of the following statements and place the corresponding letter in the space provided.

1. Periodic inspection of motors, controls, and other electrical
    equipment is important because it   _____
    a. gives advance notice of impending trouble.
    b. is required by the job standards.
    c. is a requirement of supervision.
    d. completes a day's work.

2. Careful motor troubleshooting requires the use of   _____
    a. the sense of smell   c. hearing and vision.
    b. the sense of touch   d. all of these.

3. The most accurate method of testing insulation resistance uses a(an)   _____
    a. megohmmeter.   c. ohmmeter.
    b. growler   d. tachometer.

4. Insulation resistance is measured in   _____
    a. megawatts.   c. kilohms.
    b. megohms.   d. kilovolts.

5. A very low value of insulation resistance indicates
    a. a good operating condition.
    b. a fair operating condition.
    c. an immediate investigation.
    d. that the measuring instrument is at the wrong setting.

6. Short-circuited coils are located efficiently by the use of a(an)
    a. megohmmeter.   c. ohmmeter.
    b. growler.   d. clamp-on ammeter.

7. On a balance test for phase currents where one phase shows a
    higher reading, the probable cause is   _____
    a. open turns.
    b. shorted turns.
    c. worn bearings.
    d. the need for rotor balancing.

8. If a three-phase, squirrel-cage motor is operating with current in
    only two line leads, then   _____
    a. the motor is operating as a two-phase motor.
    b. the insulation is overloaded.
    c. there is an open circuit in the stator.
    d. there is an open circuit in the rotor.

9. A properly-sized ball bearing race is mounted on a shaft correctly by tapping _____
   a. a metal tube on the outer ring.
   b. the inner ring with a hammer.
   c. the outer ring with a hammer.
   d. a metal tube on the inner ring.

10. A ball-bearing, ten-hp, three-phase, 230-volt induction motor operating under very severe conditions should be greased about every _____
    a. three months.   c. twelve months.
    b. six months.   d. month.

B. Insert the word or phrase to complete each of the following statements.

11. The instrument used to measure insulation resistance is operated with the test leads unconnected. The meter reading should be _____.

12. The measurement of individual phase currents in the operation of a three-phase induction motor is called a(an) _____.

13. The rotor and stator concentricity in three-phase motors and alternators can be determined by measurements of the _____.

14. Ball bearings should be removed from a motor shaft using a bearing _____.

15. Pressure should never be applied to the _____ race of a ball bearing.

16. Ball bearings are lubricated with _____.

17. Insufficient oil can ruin a bearing, but excessive oiling can ruin the _____.

U •N •I •T
27

SUMMARY REVIEW OF UNITS
20-26

OBJECTIVE

- To give the student an opportunity to evaluate the knowledge and understanding acquired in the study of the previous seven units.

1. List four applications for selsyn units.

   a. _____

   b. _____

   c. _____

   d. _____

2. Draw a schematic wiring diagram of a selsyn system consisting of a transmitter and one selsyn receiver unit.

3. Explain the operation of a simple selsyn system consisting of one transmitter and one selsyn receiver unit. _____

   _____

   _____

   _____

   _____

4. Explain the purpose of a differential selsyn unit. _____

_____

_____

_____

5. Draw a schematic wiring diagram of a selsyn system having one selsyn exciter unit, a differential selsyn unit, and one selsyn receiver unit.

6. Insert the correct word or phrase to complete each of the following statements.
   a. Whenever the rotor of the _____ is out of alignment with the rotor of the receiver selsyn, currents are present in the stator windings.
   b. A selsyn receiver will rotate continuously if the transmitter rotor is driven at _____ speed.
   c. The rotor units of the transmitter and receiver selsyns must be excited from the same _____ .
   d. The use of a mechanical damper on the rotor of selsyn receivers minimizes any tendency of the receiver to _____ .

7. List four applications for a split-phase induction motor.
   a. _____
   b. _____
   c. _____
   d. _____

8.   What are the basic parts of a split-phase induction motor? _____

   _____

   _____

   _____

9.   Explain how the direction of rotation of a split-phase induction motor is reversed.

   _____

   _____

   _____

10.   What happens if the centrifugal switch contacts fail to reclose when a split-phase
   motor is stopped? _____

   _____

11.   A split-phase induction motor has a dual-voltage rating of 115/230 volts. The motor
   has two running windings, each of which is rated at II 5 volts. The motor also has
   two starting windings, each of which is rated at I *1* 5 volts. Draw a schematic connection diagram of this split-phase induction motor connected for 230 volts.

12.   What is the basic difference between a split-phase induction motor and a capacitor
   start, induction run motor? _____

   _____

   _____

   _____

   _____

13. If the centrifugal switch fails to open as a split-phase motor accelerates to the rated speed, what happens to the starting winding? _____

_____

_____

14. If the centrifugal switch on a capacitor start, induction run motor fails to open as the motor accelerates to the rated speed, what may happen in the starting winding circuit?

_____

_____

15. What is one limitation of a capacitor start, induction run motor? _____

_____

_____

16. What is the basic difference between a capacitor start, induction run motor and a capacitor start, capacitor run motor? _____

_____

_____

_____

_____

17. List three types of capacitor start, capacitor run motors.
    a. _____
    b. _____
    c. _____

18. Insert the correct word or phrase to complete each of the following statements.
    a. The capacitor in series with the starting winding of a capacitor start, induction run motor improves the _____ torque of the motor.
    b. A split-phase induction motor has good speed regulation but _____ starting torque characteristics.
    c. A capacitor start, capacitor run motor has practically _____ power factor when operating at full load.
    d. A capacitor start, capacitor run motor has _____ starting torque and _____ speed regulation.
    e. A capacitor start, induction run motor has _____ speed regulation.

19. Insert the correct word or phrase to complete each of the following statements.
    a. The capacitor used with a capacitor start, induction run motor is used only for the purpose of improving the _____ of the motor.
    b. The capacitors used with a capacitor start, capacitor run motor are used to improve _____ .

   c. A motor of one horsepower or less which is manually started and which is within sight of the starter location, provided the distance is no greater than 50 feet, is considered protected by the _____ .

   d. A motor of one horsepower or less which is manually operated but more than 50 feet from the starter location shall have a _____ _____ .

   e. Where separate overcurrent devices are required for motors, they shall not be set at more than _____ percent of the motor nameplate full-load current rating for motors marked to have a temperature rise not over _____ for motors with a marked service factor of 1.15, and at not more than _____ percent for all other types of motors.

20. Where is a repulsion motor used? _____

    _____

    _____

    _____

21. What are the two types of repulsion start, induction run motors?

   a._____

   b._____

22. Where are repulsion start, induction run motors used? _____

    _____

    _____

23. A 3-hp, 230-volt, 17-ampere, single-phase repulsion start, induction run motor is connected directly across rated line voltage.

   a. Determine the overcurrent protection for the branch circuit feeding this motor.

    _____

    _____

   b. Determine the running overcurrent protection to use with this motor.

    _____

    _____

24. What size copper wire (Type THHN) is used for the branch circuit feeding the motor in question 23? _____

    _____

    _____

25. What is a repulsion-induction motor? _____

    _____

    _____

    _____

26. What is one advantage to the use of the repulsion-induction motor as compared with the repulsion start, induction run motor?_____

_____

_____

27. Insert the correct word or phrase to complete each of the following statements.
    a. A repulsion motor has good _____ but poor _____.
    b. The speed of a repulsion motor can be controlled by changing the _____.
    c. Both the brush-riding and the brush-lifting types of repulsion start, induction after they have run motors operate as _____ after they have accelerated to rated speed.
    d. The repulsion-induction motor has good _____ and relatively good
    _____.

28. Explain how the direction of rotation is changed on any one of the three types of single-phase repulsion motors._____

_____

_____

29. What is a universal motor? _____

_____

_____

30. Draw a schematic diagram of a conductively compensated series motor.

31. Draw a schematic diagram of an inductively compensated series motor.

32. In what way is an inductively compensated series motor different from a conductively compensated series motor? _____

_____

_____

_____

33. Explain how the direction of rotation is reversed for a conductively compensated series motor._____

_____

_____

34. What is the purpose of a compensating winding in an ac series motor? _____

_____

_____

_____

_____

35. A universal motor can be operated on   _____
    a. ac power only.
    b. dc power only.
    c. ac or dc power.

36. A conductively compensated series motor can be operated on   _____
    a. ac power only.
    b. dc power only.
    c. ac or dc power.

37. An inductively compensated series motor can be operated on either
    ac or dc power.   _____
    a. true
    b. false

38. A large 25-hp, direct-current series motor will not operate satis-
    factorily on an ac power source.   _____
    a. true
    b. false

GLOSSARY

ACROSS-THE-LINE. Method of motor starting which connects the motor directly to the supply line, on starting or running; also called *full voltage control*.

ALTERNATING CURRENT (ac). A current which alternates regularly in direction. Refers to a periodic current with successive half waves of the same shape and area.

ALTERNATOR. A machine used to generate alternating current by rotating conductors through a magnetic field; an alternating current generator.

ALTERNATOR PERIODIC TIME RELATIONSHIP. The phase voltages of two generators running at different speeds.

ALTERNATORS PARALLELED. Alternators are connected in parallel whenever the power demand of the load circuit is greater than the power output of a single alternator.

AMBIENT TEMPERATURE. The temperature surrounding a device.

AMORTISSEUR WINDING. Consists of copper bars embedded in the cores of the poles of a synchronous motor. The copper bars of this special type of squirrel-cage winding are welded to end rings on each side of the rotor; used for starting only.

ARMATURE. A cylindrical, laminated iron structure mounted on a drive shaft; contains the armature winding.

ARMATURE WINDING. Wiring embedded in slots on the surface of the armature; voltage is induced in this winding on a generator.

AUTOMATIC COMPENSATORS. Motor starters that have provisions for connecting three-phase motors automatically across 50%, 65%, 80%, and 100% of the rated line voltage for starting, in that order after preset timing.

AUTOTRANSFORMER. A transformer in which a part of the winding is common to both the primary and secondary circuits.

AUXILIARY CONTACTS. Contacts of a switching device in addition to the main circuit contacts; auxiliary contacts operate with the movement of the main contacts; electrical interlocks.

BLOWOUT COIL. Electromagnetic coil used in contactors and motor starters to deflect an arc when a circuit is interrupted.

BRANCH CIRCUIT. That portion of a wiring system that extends beyond the final overcurrent device protecting the circuit.

BRUSHLESS EXCITATION. The commutator of a conventional direct-connected exciter of a synchronous motor is replaced with a three-phase, bridge-type, solid-state rectifier.

BRUSHLESS EXCITER. Solid-state voltage control on an alternator, providing dc necessary for the generation of ac.

BUS. A conducting bar, of different current capacities, usually made of copper or aluminum.

BUSWAY. A system of enclosed power transmission that is current and voltage rated.

CAPACITOR. A device made with two conductive plates separated by an insulator or dielectric.

CENTRIFUGAL SWITCH. On single-phase motors, when the rotor is at normal speed, centrifugal force set up in the switch mechanism causes the collar to move and allows switch contacts to open, removing starting winding.

CIRCUIT BREAKER. A device designed to open and close a circuit by nonautomatic means and to open the circuit automatically on a predetermined overcurrent without injury to itself when properly applied within its rating.

COGENERATING PLANTS. Diesel powered electric generating sets which are designed to recapture and use the waste heat both from their exhaust and cooling systems.

COMMUTATOR. Consists of a series of copper segments which are insulated from one another and the mounting shaft; used on dc motors and generators.

COMPENSATOR TRANSFORMER. A tapped autotransformer which is used for starting induction motors.

CONDUCTOR. A device or material that permits current to flow through it easily.

CONDUIT PLAN. A diagram of all external wiring between isolated panels and electrical equipment.

CONTACTOR. An electromagnetic device that repeatedly establishes or interrupts an electric power circuit.

CONTROLLER. A device or group of devices that governs, in a predetermined manner, the delivery of electric power to apparatus connected to it.

COUNTER EMF. An induced voltage developed in a dc motor while rotating. The direction of the induced voltage is opposite to that of the applied voltage.

CUMULATIVE COMPOUND-WOUND GENERATOR OR MOTOR. A series winding is connected to aid the shunt winding.

CURRENT. The rate of flow of electrons which is measured in amperes.

CURRENT FLOW. The flow of electrons.

DC EXCITER BUS. A bus from which other alternators receive their excitation power.

DEFINITE TIME. A predetermined time lapse.

DELTA CONNECTION. A circuit formed by connecting three electrical devices in series to form a closed loop; used in three-phase connections.

DIODE. A two-element device that permits current to flow through it in only one direction.

DIRECT CURRENT (dc). Current that does not reverse its direction of flow. It is a continuous nonvarying current in one direction.

DISCONNECTING SWITCH. A switch which is intended to open a circuit only after the load has been thrown off by some other means, not intended to be opened under load.

DRUM SWITCH. A manually operated switch having electrical connecting parts in the form of fingers held by spring pressure against contact segments or surfaces on the periphery of a rotating cylinder or sector.

DUAL VOLTAGE MOTORS. Motors designed to operate on two different voltage ratings.

DUTY CYCLE. The period of time in which a motor can safely operate under a load. *Continuous* means that the motor can operate fully loaded 24 hours a day.

DYNAMIC BRAKING. Using a dc motor as a generator, taking it off the supply line and applying an energy dissipating resistor to the armature. Dynamic braking for an ac motor is accomplished by disconnecting the motor from the line and connecting dc power to the stator windings.

EDDY CURRENT.   Current induced into the core of a magnetic device. Causes part of the iron core losses, in the form of heat.

EFFICIENCY.   The efficiency of all machinery is the ratio of the output to the input.

$$\frac{\text{output}}{\text{input}} = \text{efficiency}$$

ELECTRIC CONTROLLER.   A device, or group of devices, which governs, in some predetermined manner, the electric power delivered to the apparatus to which it is connected.

ELEMENTARY DIAGRAM (Ladder Diagram, Schematic Diagram, Line Diagram).   Represents the electrical control circuit in the simplest manner. All control devices and connections are shown as symbols located between vertical lines that represent the source of control power.

EMERGENCY GENERATOR SYSTEM.   A generating set which functions as a power source in a health care facility, such as a hospital; a standby power system. In addition to lighting, the loads supplied are essential to life and safety.

ENGINE-DRIVEN GENERATING SETS.   Generators with prime movers of diesel or gasoline engines, or natural gas, and the like.

EXCITER.   A dc generator that supplies the magnetic field for an alternator.

FEEDER.   The circuit conductor between the service equipment or the switchboard of an isolated plant and the branch-circuit overcurrent device.

FIELD DISCHARGE SWITCH.   Used in the excitation circuit of an alternator. Controls (through a resistor) the high inductive voltage created in the field coils by the collapsing magnetic field.

FLUX.   Magnetic field; lines of force around a magnet.

FREQUENCY.   Cycles per second or hertz.

FUSE.   An overcurrent protective device with a circuit opening fusible part that is heated and severed by the passage of overcurrent through it.

GEAR MOTOR.   A self-contained drive made up of a ball bearing motor and a speed reducing gear box.

GROUNDED.   Connected to earth or to some conducting body that serves in place of earth.

GROWLER.   An instrument consisting of an electromagnetic yoke and winding excited from an ac source; used to locate short-circuited motor coils.

HERTZ.   The measurement of the number of cycles of an alternating current or voltage completed in one second.

HYSTERESIS.   Part of iron core losses.

IDENTIFIED CONDUCTOR (Neutral).   A grounded conductor in an electrical system, identified with the code color white.

INDUCED CURRENT.   Current produced in a conductor by the cutting action of a magnetic field.

INDUCED VOLTAGE.   Voltage created in a conductor when the conductor interacts with a magnetic field.

INDUCTION.   Induced voltage is always in such a direction as to oppose the force producing it.

INSULATOR.   Material with a very high resistance which is used to electrically isolate two conductive surfaces.

I/O SECTION–INPUT/OUTPUT SECTION.   This section of a programmable controller interfaces the PC to the electrical signals in the field. The input takes the appropriate field indication and converts it to a signal recognizable by the processor. The output takes a signal sent by the microprocessor and converts it to the proper signal for the field devices.

ISOLATING TRANSFORMER.   A transformer in which the secondary winding is electrically isolated from the primary winding.

JOGGING.   The quickly repeated closure of a controller circuit to start a motor from rest for the purpose of accomplishing small movements of the driven machine.

LEGALLY REQUIRED STANDBY GENERATING SYSTEMS.   Those systems required by municipal, state, federal or other codes or government agency having jurisdiction.

LENZ'S LAW.   A voltage is induced in a coil whenever the coil circuit is opened or closed.

MAINTAINING CONTACT.   A small contact in the control circuit used to keep a coil energized, usually actuated by the same coil; also known as a holding contact or an auxiliary contact.

MECHANICAL INTERLOCK.   A mechanical interlocking device is assembled at the factory between forward and reverse motor starters and multispeed starters; it locks out one starter at the beginning of the stroke of either starter to prevent short circuits and burnouts by the accidental closure of both starters simultaneously.

MEGOHMMETER (MEGGER®).   An electrical instrument used to measure insulation resistance.

MEGOHMS.   A unit of resistance equal to 1,000,000 ohms.

MOTOR CIRCUIT SWITCH (Externally Operated Disconnect Switch, EXO).   Motor branch circuit switch rated in horsepower. Usually contains motor starting protection; safety switch.

MOTOR CONTROLLER.   A device used to control the operation of a motor.

MOTORIZING.   A generator armature rotates as a motor.

MOTOR STARTER.   A device used to start and/or regulate the current to a motor during the starting period. It may be used to make or break the circuit and/or limit the starting current. It is equipped with overload protection devices, such as a contactor with overload relays.

MULTIMETER.   Electrical instrument designed to measure two or more electrical quantities.

NEC.   National Electrical Code.

NONSALIENT ROTOR.   A rotor that has a smooth cylindrical surface. The field poles (usually two or four) do not protrude above this smooth surface.

NORMAL FIELD EXCITATION.   The value of dc field excitation required to achieve unity power factor in a synchronous motor.

NORMALLY OPEN and NORMALLY CLOSED.   When applied to a magnetically operated switching device, such as a contactor or relay, or to the contacts of these devices, these terms signify the position taken when the operating magnet is deenergized, and with no external forces applied. The terms apply only to nonlatching types of devices.

OHMMETER.   An instrument used to measure resistance.

OIL (IMMERSED) SWITCH.   Contacts of a switch that operate in an oil bath tank. Switch is used on high voltages to connect or disconnect a load. Also known as an oil circuit breaker.

OVERLOAD.   Operation of equipment in excess of normal, full load rating, or of a conductor in excess of rated ampacity which, when it persists for a sufficient length of time, would cause damage or dangerous overheating.

OVERLOAD PROTECTION (Running Protection).   Overload protection is the result of a device that operates on excessive current, but not necessarily on a short circuit, to cause the interruption of current flow to the device governed.

PARALLEL CIRCUIT.   A circuit that has more than one path for current flow.

PERMEABILITY.   The ease with which a material will conduct magnetic lines of force.

PLUGGING.   Braking a motor by reversing the line voltage or phase sequence; motor develops a retarding force; a quick stop.

PLUGGING RELAY.   A device attached to a motor shaft to accomplish plugging switches reversing starter to establish counter torque which brings the motor to a quick standstill before it begins to rotate in the reverse direction.

POLARITY.   Characteristic (negative or positive) of a charge. The characteristic of a device that exhibits opposite quantities, such as positive and negative, within itself.

POLE.   The north or south magnetic end of a magnet; a terminal of a switch; one set of contacts for one circuit of main power.

POLYPHASE.   An electrical system with the proper combination of two or more single-phase systems.

POWER FACTOR.   The ratio of true power to apparent power. A power factor of 100% is the best electrical system.

PREVENTIVE MAINTENANCE.   Periodic inspections to prevent serious damage to machinery by locating potential trouble areas; preventing breakdowns rather than repairing them.

PROCESSOR.   The microprocessor section of a programmable controller. It is the section of a PC that holds the programs, receives information, makes a decision and delivers an output signal to some external electrical device.

PROGRAMMABLE CONTROLLER (PC).   A microprocessor based system used to control electrical operations. It is a control system controlled by software that can easily be altered to provide flexible control schemes.

PROGRAMMER.   A device–either hand held, or personal computer, or special monitor and keyboard–that allows a person to enter desired control programs to the microprocessor section of a PC.

PULSE WIDTH MODULATION (PWM).   A process that controls the width of a pulse delivered to an ac motor. By modulating the width of several pulses and also controlling the amplitude, a waveform that approximates a sine wave is produced at adjustable frequencies.

PUSH BUTTON.   A master switch; manually operated plunger or button for an electrical actuating device; assembled into push-button stations.

RACEWAY.   A channel, or conduit, designed expressly for holding wires, cables, or busbars.

RATING.   The rating of a switch or circuit breaker includes (1) the maximum current and voltage of the circuit on which it is intended to operate, (2) the normal frequency of the current; and (3) the interrupting tolerance of the device.

RECTIFIER.   A device that converts alternating current (ac) into direct current (dc).

REGULATION. Voltage at the terminals of a generator or transformer, for different values of the load current; usually expressed as a percentage.

RELAY. Used in control circuits; operated by a change in one electrical circuit to control a device in the same circuit or another circuit.

REMOTE CONTROL. Controls the function initiation or change of an electrical device from some remote place or location.

RESIDUAL FLUX. A small amount of magnetic field.

RESISTANCE STARTER (Primary Resistance Starter). A controller to start a motor at a reduced voltage with resistors in the line on start.

RHEOSTAT. A resistor that can be adjusted to vary its resistance without opening the circuit in which it may be connected.

R/min (or RPM). Speed in revolutions per minute.

ROTOR. The revolving part of an ac motor or alternator.

SALIENT FIELD ROTOR. Found on three-phase alternators and synchronous motors; field poles protrude from the rotor support structure. The structure is of steel construction and commonly consists of a hub, spokes, and a rim. This support structure is called a spider.

SELSYN. Abbreviation of the words self-synchronous. Selsyn units are special ac motors used primarily in applications requiring remote control. These units are also referred to as synchros.

SEMICONDUCTORS. Materials which are neither good conductors not good insulators. Certain combinations of these materials allow current to flow in one direction but not in the opposite direction.

SEPARATE CONTROL. The coil voltages of a relay, contactor or motor starter are separate or different from those at the switch contacts.

SEPARATELY-EXCITED FIELD. The electrical power required by the field circuit of a dc generator may be supplied from a separate or outside dc supply.

SERIES FIELD. In a dc motor, has comparatively few turns of wire of a size that will permit it to carry the full load current of the motor.

SERIES WINDING. Generator winding connected in series with the armature and load; carries full load.

SERVICE FACTOR. An allowable motor overload; the amount of allowable overload is indicated by a multiplier which, when applied to a normal horsepower rating, indicates the permissible loading.

SHORT AND GROUND. A flexible cable with clamps on both ends. It is used to ground and short high lines to prevent electrical shock to workmen.

SHUNT. To connect in parallel; to divert or be diverted by a shunt.

SILICON-CONTROLLED RECTIFIER (SCR). A four-layer semiconductor device that is a rectifier. It must be triggered by a pulse applied to the gate before it will conduct electricity.

SINGLE-PHASE. A term characterizing a circuit energized by a single alternating emf. Such a circuit is usually supplied through two wires.

SLIP. In an induction motor, slip is the difference between the synchronous speed and the rotor speed, usually expressed as a percentage.

SLIP RINGS. Copper or brass rings mounted on, and insulated from, the shaft of an alternator or wound rotor induction motor; used to complete connections between a stationary circuit and a revolving circuit.

SOLENOID. An electromagnet used to cause mechanical movement of an armature, such as a solenoid valve.

SOLID STATE. As used in electrical-electronic circuits, refers to the use of solid materials as opposed to gases, as in an electron tube. It usually refers to equipment using semiconductors.

SPEED CONTROL. Refers to changes in motor speed produced intentionally by the use of auxiliary control, such as a field rheostat or automatic equipment.

SPEED REGULATION. Refers to the changes in speed produced by changes within the motor due to a load applied to the shaft.

SPLIT PHASE. A single-phase induction motor with auxiliary winding, displaced in magnetic position from, and connected parallel to, the main winding.

STANDBY POWER GENERATING SYSTEM. Alternate power system for applications such as heating, refrigeration, data processing, or communications systems where interruption of normal power would cause human discomfort or damage to a product in manufacture.

STARTING CURRENT. The surge of amperes of a motor upon starting.

STARTING PROTECTION. Overcurrent protection is provided to protect the motor installation from potential damage due to short circuits, defective wiring, or faults in the motor controller or the motor windings. The starting protection may consist of a motor disconnect switch containing fuses.

STATOR. The stationary part of a motor or alternator; the part of the machine that is secured to the frame.

SYNCHRONOUS ALTERNATORS. Frequencies, voltages, and instantaneous ac polarities must be equal when paralleling machines.

SYNCHRONOUS CAPACITOR. A synchronous motor operating only to correct the power factor and not driving any mechanical load.

SYNCHRONOUS MOTOR. A three-phase motor (ac) which operates at a constant speed from a no load condition to full load; has a revolving field which is separately excited from a direct current source; similar in construction to a three-phase ac alternator.

SYNCHRONOUS SPEED. The speed at which the electromagnetic field revolves around the stator of an induction motor. The synchronous speed is determined by the frequency (hertz) of the supply voltage and the number of poles on the motor stator.

SYNCHROSCOPE. An electrical instrument for synchronizing two alternators.

TACHOMETER. An instrument used to check the speed of a motor or machine.

THREE PHASE. A term applied to three alternating currents or voltages of the same frequency, type of wave, and amplitude. The currents and/or voltages are one third of a cycle (120 electrical time degrees) apart.

THREE-PHASE SYSTEM. Electrical energy originates from an alternator which has three main windings placed 120 degrees apart. Three wires are used to transmit the energy.

THYRISTOR. An electronic component that has only two states of operation–on or off.

TORQUE. The rotating force of a motor shaft produced by the interaction of the magnetic fields of the armature and the field poles.

TRANSFER SWITCHES. Switches to transfer, or reconnect, the load from a preferred or normal electric power supply to the emergency power supply.

TRANSFORMER. An electromagnetic device that converts voltages for use in power transmission and operation of control devices.

TRANSFORMER BANK.   When two or three transformers are used to step down or step up voltage on a three-phase system.

TRANSFORMER SECONDARY WINDING.   The coil that discharges the energy at a transformed or changed voltage, up or down.

WHEATSTONE BRIDGE.   Circuit configuration used to measure electrical qualities such as resistance.

WIRING DIAGRAM.   Locates the wiring on a control panel in relationship to the actual location of the equipment and terminals; made up of a method of lines and symbols on paper.

WOUND ROTOR INDUCTION MOTOR.   An ac motor consisting of a stator core with a three-phase winding, a wound rotor with slip rings, brushes and brush holders, and two end shields to house the bearings that support the rotor shaft.

WIRING DIAGRAM.   Locates the wiring on a control panel in relationship to the actual location of the equipment and terminals; specific lines and symbols represent components and wiring.

WYE CONNECTION (Star).   A connection of three components made in such a manner that one end of each component is connected. This connection generally connects devices to a three-phase power system.

INDEX